PSYCHOLOGY RESEARCH PROGRESS

MIND OVER MATTER

Psychology Research Progress

Additional books in this series can be found on Nova's website under the Series tab.

Additional e-books in this series can be found on Nova's website under the eBooks tab.

PSYCHOLOGY RESEARCH PROGRESS

MIND OVER MATTER

LUKE STRONGMAN

nova publishers
New York

NOTICE TO THE READER

Library of Congress Cataloging-in-Publication Data

ISBN: 978-1-63485-739-0

Published by Nova Science Publishers, Inc. † New York

CONTENTS

INTRODUCTION

Mind over Matter explores issues in popular psychology and science that are firstly concerned with the evolution of the human species and, more particularly, of the human mind and its functioning; secondly, it discusses how people orientate to their environments and react through communication and discourse. It will also explore some interesting bisecting roads of neural functioning, such as the purpose of déjà vu and dreaming, discuss what happens when the mind seeks "shortcuts" in the creation of understanding, and what happens at the periphery of conscious control. *Mind over Matter* is about the human mind and making sense of the semiosphere of information that surrounds us. As such, it is both about something – mind and behaviour – and also a reflection on thinking about the mind. The book is concerned with thinking through the ideas behind orientating oneself in an environment which is teeming with information, consciousness, and the possibilities of language and communication. *Mind over Matter* concludes by making brief commentary on possible human evolution in the future.

As such the scope of *Mind over Matter,* like a bridge, is both broad and narrow.

Chapter One sets the scene for the genome generation, the first human generation in the twenty-first century to be born with knowledge of the genome sequence. Being able to map, understand, and ultimately 'manipulate' the human genome will bring about vast changes to the way we think about human life, how we store personal information, how medical treatments are devised according to genetic compositions and also possibly to the future of human beings – so-called 'designer babies' -- those whose parents 'choose' some components of their genetic make-up. Chapter Two surveys developments in evolutionary psychology that spans human development from the first tool use some four million, five-hundred thousand years ago, which many anthropologists believe is the defining quality of a human-like characteristic, to explore the time-frames over which people are known to evolve and the limitation this places on understanding the development of the mind. The discussion is derived from the product of two sources of knowledge. On the one hand we have fossil evidence to delineate human development and on the other hand we have knowledge of genomics and predictive rationales based on psych-social inference for explaining likely evolutionary trends in the ongoing natural selection of human cognitive ontogeny and phylogeny. Chapter Three looks more closely at the way the human functions in the environment, and the idea of the human being a product of the environment and always being in relation to it, to the development of the inter-relationship between place and mind, of habitus and habituation as influences on human behaviour within a semiosphere of environmental information. Chapter Four extends concepts of evolution and place relatedness to discuss camouflage from an organicist perspective, to begin to see the way in which the human mind can both consciously and unconsciously manipulate the environment to reveal or conceal its intent, much as though predator and prey in the animal kingdom. Camouflage is seen as a property of consciousness, a conscious manipulation of environmental conditions rather than something that is hard-wired into the form of an organism, though such 'hard-wrings' are possible and do indeed occur. A chameleon will change its colour, automatically, in response to environmental conditions, much as though a person might change thinking states, moods and dispositions according to emotions. The language of the animal kingdom is physical and functional whereas human language is metaphysical and symbolic. However, it can be seen that human symbolic manipulation and language use, the generative grammar that enables controlled thought and the creation of self-awareness and consciousness nevertheless mimics a process that is hard-wired between an organism and its environment. People have an innate capacity for generative grammar, we are not born speaking but are born with the capacity and propensity for language ready-made. Chapter Five examines language evolution and looks at how language has derived from culture and what forms it takes. Chapter Six is a study of a by-product of physiological arousal that allows a glimpse into the generation of cognitive states of adjustment – the phenomenon of déjà vu. The argument is made that déjà vu is an experience of modernity and of the mind undergoing susceptibility to change, as such it may signal a process of evolutional adaptation at physiological in cognate and cortex functioning. Chapter Seven examines the patterns within the complex boundaries between states of matter, it describes conditions under which hidden order may be revealed from chaotic pattern and how science may be understood from the perspective of an anti-reductionist viewpoint. Chapter Eight involves a discussion of the ways in which the mind may fall itself in attempting to process information and understand as efficiency as possible the complexity of information offered to it by language, human interchange and in the semiosphere of meaning which we all inhabit and create for ourselves. Chapter Nine examines the theory of cooperation and competition to the twin propensities that govern human behaviours according philosophies of social Darwinism. Chapter Ten looks more closely at the effects of language interchange, techniques of cognitive interviewing and the psychology of disclosure probing at the ways in which while concepts of the human may involve greater socialisation because of developments in societal structure and information and computer technologies (ICT) technologies, nevertheless the mind invents strategies of processing and information recall that link the devices of memory to context. Disclosure reveals the economies of self-knowledge, deceit and the possibilities for more effective relations between people. Chapter Eleven is concerned with the personal and the interpersonal, and the emotions of confidence and embarrassment that govern our special lives and our self-perceptions. Understanding these emotions may be the root of maintain effective relationships and useful detachment in the everyday world. Chapter Twelve is about the processes of human dreaming, its function and states, and how sleep is also related to information recall and memory. Chapter Thirteen discusses the phenomenon of auditory hallucination as existing on a continuum of non-pathological behaviours that may be concerned with human cognitive development. Chapter Fourteen looks briefly at the phenomenon of resilience which has become the catch-cry of modern living, the response to adversity and the ability to keep functioning robustly in trying conditions. Correspondingly this chapter also looks briefly at the psychology of human security, a desired state of being, and how it might be obtained the face of a challenging environment. Finally, a possible future for human evolutionary development is sketched, and so too the affirmation that the human species has not yet reached the fullness of its adaptive potential, particularly in response to the recent technological innovations and the socio-political economic organisation of modern society and the extent for the new found human control over the environment.

Chapter 1

THE GENOME GENERATION

People born in the twenty-first century will be the first generation to be born in a society which has ever-increasing knowledge of the human genome (especially among research scientists and medical professionals), despite its inherent complexity. Currently science has a map of the human genome, but the full implication between the relationships of constituent parts is not yet fully understood. The human species is subject to age-old evolutionary process that in part constitutes our 'destiny' as a species and over which we do not and perhaps will never have complete control. All living things are part of this inexorable phenomenon. We are evolving and always have been – we are subject to a process which paradoxically we help make but which is not of our making in a conscious sense. Evolution and natural selection 'designs' the human body for a purpose as a product of the environment, and so it is with all other living organisms. We perform an ecological function despite of and perhaps especially because of, our unprecedented control over it as a society, through harnessing agriculture, industry and combustion and biological processes, to name just a few areas of technological development. Evolution tends to adapt for functionality, and that which is not used is eventually lost.

The way in which evolution (or the unfolding) occurs is by natural selection through transfer of DNA between the generations of a species.[1]

[1] DNA stands for deoxyribonucleic acid and it is a molecule which carries genetic instructions from one generation to another. These genetic instructions are used in development functioning and reproduction of organisms and consist of two bi-polymer strands coiled to make a double-helix of polynucleotides consisting of a nucleobase of either cystosine (C), guanine (G), ademine (A), or thymine (T) which a nitogen based and a deoxyribose sugar which join together in bonds between nucleotide sugars and phosphate according to base-

The genetic material of an organism is called the genome[2] which consists of either DNA or RNA[3], the genes including coding and non-coding[4] regions and the mitochondria and chloroplasts. We have approximately twenty-thousand genes which is not as high as some other life-forms considered rudimentary in comparison. Approximately five-per cent of the genome differentiates humans from other primates. That part of the genome that is intergenerative coding is possibly only three-per cent. With the comparatively recent mapping of the human genome, the role of so-called 'junk-DNA' hitherto thought 'inactive' is increasingly shown to have a bearing of phenotype expression. Hence in our genetic make-up the redundancy of so-called 'junk-genes' or non-coding genes has been brought into question by new knowledge of the human genome. The portion of our genetic make-up that has a biochemical function is thought to be larger than that ascribed to it prior to knowledge of genome sequencing, some eighty-per cent as opposed to the two-per cent hitherto known to encode protein sequences.

Dawkins (2006) suggests only one-per cent of total *information* in human cells appears to be used, the purpose of the other ninety-nine per cent is unknown [1]. However, knowledge of this one-per cent is potentially fraught with social, ethical and legal consequences because with new technological developments it involves the scientific manipulation of the genetic code which enables human life and raises issues to do with the modification of it. If as a parent you had a chance to screen an unborn child for syndrome X would you do it, knowing that a positive result would mean a new set of ethical choices? What if you could select trait Y over trait Z in 'designing your child's genetic make-up, does such control hitherto absent from the genetic 'lottery' of life, impose psycho-social consequences on parenting? If the pressure for genetic

pairing (A with T and C with G). Friedrich Miescher first isolated DNA in 1869 and its molecular structure was identified by James Watson and Francis 1953.

[2] The term genome was invented by Hans Winkler in 1920 – by combing the words gene and chromosome.

[3] Ribonucleic acid (RNA) has a function of coding, decoding, regulating and expressing genes. It is comprised of a chain of necleotides as a single-strand folded onto itself and plays a role in catalysing biological reactions, sensing and communicating responses to cellular signals and controlling gene expression. RNA is involved in protein synthesis whereby they regulate the composition of proteins on ribosomes via tRNA (or transfer RNA molecules which deliver amino acids to a ribosome and then link these together forming proteins.

[4] Noncoding DNA sequences are those that don't encode protein sequences. Noncoding DNA is transcribed into functional noncoding RNA (with transfer, ribosomal and regulatory functioning), and / or regulation of protein-coding sequences, scaffolding attachment, and in centromeres and telomers. It is thought that at least eighty per cent of the human genome has biochemical activity (ENCODE), the functional fraction of which is between eight and fifteen per cent.

modification or selection were to produce an offspring whose genes had been consciously modified, what unknown effects might there be on the child's future and the future of his or her children? Such a person would truly be a 'stranger in a strange land.'

Despite the fact that as Dawkins (2006) points out there is enough information capacity in a single cell to "store the Encyclopaedia Britannica, all 30 volumes of it, three or four times over" manipulating this code is not like editing a book because it involves manipulating the variables that are at the core of human being – sanctified by ethical norms and legal codes created by people who for the most part have come into existence by 'natural' pro-creation [2]. If we can 'design' our children's genes rather than let 'mother nature' do it for us, this causes us reconsider such notions as what does it mean to be human? Does altering the odds in the genetic lottery of life, cause 'downstream effects in the gene pool that we cannot foresee'? Will we thereby 'breed-out' items of junk DNA before we truly know the functions of it? Do we really have the knowledge to tinker with a system that has been millions of years in the creating? Does the manner of creation impose limitations on the creator and the created? One human being is no more or less human than another in the current grand scheme of life, altering the genetic make-up of our children might be to create a sub-species of 'other humans' that would be counter to the genetic homogenizing tendencies in our current global culture. Would belonging to a 'new breed' of human being be too perilous a fate? While people of very similar genetic make-up may have different personalities, arguably human existence is defined by qualities of experience and perception, the qualia of thoughts, feelings, sensations and memories – and these qualities are not reducible to genetic quantification. Designing peoples' genetic make-up is still largely in the realm of science- fiction if only because no government is willing to set aside the ethical argument against it. But how long will it remain so?

Perhaps research into the human genome and bio-medical applications will allow society to eliminate genetic diseases before the creation genetically engineered human beings is made possible. Here research is more likely to point to therapies for single gene diseases but polygenic diseases which have unknown inter-relationships between genes are far more complex and will require much more research before scientists have finally mapped them in the genome or understood their characteristics in phenotype expression. Though this area of research shows great potential in medical research whereby therapies can be designed for a patient according to their genetic make-up. Knowledge of the genome has also brought gene editing technologies which uses molecular scissors to 'edit' DNA strands enabling their insertion, deletion and repair. Arguably also, overcoming sterility is just one step. Genetic research is increasingly geared towards knowing genome maps of children that are yet to be born – which involves ethical dilemmas of freedom of another kind. The argument both for and against the genetic modification of humans is based around deontological and rights based ethics, the issue of harming people by depriving them of some kinds of ontological freedoms, for example, versus granting freedoms through the ability to extend life. Despite genome research, biological science still doesn't know enough to ensure that children produced from genetic engineering would not suffer harm (either from their make-up) or how society treats them as products of science rather than procreation. The sociology is unknown.

If genetic research cures suffering such as with development of somatic gene therapy, is then society obligated to allow science and medicine to develop such therapies? If the relief of human suffering is the outcome it is not, then offence to human dignity and shouldn't be an obstacle to rights-based ethics [3]. However, germ-line gene therapy, modifying genes passed on to future generations is not currently performed on humans because there is a consensus against its use [4].

The argument of freedom and autonomy is two-sided however. The argument of freedom to create human beings and freedom to relive suffering is possibly as strong as the argument for negative freedoms such as the right not to interfere with 'natural' processes of creation. Furthermore, as de Melo-Martin (2005) suggests, "the flood of arguments against human cloning on grounds of a possible loss of a sense of individuality or unique identity promote the incorrect belief that our genes determine human individuality or identity" [5]. The latter arguments are most likely to be undermined by deontological and non-consequentialist ethics. The principle of reproductive freedom to create children of choice, is a powerful one and the protection of such creation is an objective aim [6]. However, counter to this is as Rodota (2002) suggests, "[c]loning marks the waiver of sexual reproduction; it challenges the uniqueness of individuals and makes human 'serialisation' a reality; it is also a manifesto of male superficiality" [7]. Implicitly, this describes nothing-less than the corporatisation of the human species and the reversal of the subject-object relationship. In summary, de Melo-Martin (2005) claims that genetic determinism is problematic for three main reasons: Firstly, an empirical foundation for the belief that genes alone determine human traits and behaviours is lacking; secondly what evidence is there to say that genetic interventions rather than preventative measures are effective; thirdly, how desirable is it for people to see their genetic makeup as their fate? [8]. It is the quality of thought and experience that is valuable despite the tendency of society to argue that in almost all measures quantity counts. We have subjectivity so that amongst other things, we can make our own choices and choose to count or discount blind objectivity but also to know when to choose the path of objectivity over individual preference. As personalities whose brains are capable of producing both subjective and objective thoughts, we are creatures of paradox in many ways, yet our parents did not 'design' us objectively. What if future parents did design their children so?

Under an understanding of freedom of self-determination, it is difficult to prohibit in theory, reproductive cloning, unless that process is fundamentally counter to notions of human dignity. *Article 13* of the *Council of Europe: Convention on Human Rights and Biomedicine* confirms that, "[t]he human body and its parts shall not as such, give rise to financial gain" [9]. The abuses of this fundamental principle are too manifold to list but the ethical dimension along which they may be compromised is clearly utilitarian gain. Might a Rawlsian ethics might apply to prohibitions on genetic engineering [10]. Could the 'veil of ignorance' principle overcome the plethora of opinion and moral ambiguity of technocratic approaches to science and medicine? Despite the best intentions of an ethically aware populous, arguably, it is simply not possible to arrive at a prior position on the integrity of the human body and the intervention of science in the health of its functions that all might agree on. What we can agree on is that human life is good and that we ought to take as a society every prevention to sustain it.

More far-reaching in its implications in the larger scientific sense is the idea that the mapping of the human genome and subsequent genetic engineering might radically alter the process of 'natural selection' itself. As Dawkins (1986/2006) suggests, "[e]volution by natural selection could not be faster than the mutation rate, for mutation is, ultimately, the only way in which new variation enters the species. All that natural selection can do is accept certain new variations, and reject others. The mutation rate is bound to place an upper limit on the rate at which evolution can proceed. As a matter of fact, most of natural selection is concerned with preventing evolutionary change rather than with driving it" [11]. As such the accelerated processes of genetic engineering that will inevitably follow from the research in the mapping of the human genome will occur at a rate of progress that is in evolutionary terms is nothing less than revolutionary.

The completion of the thirteen-year human genome project (HGP) in 2003 in which American British, Japanese, Chinese, French, German and other international scientists succeeded in identifying the approximately twenty to twenty-five thousand genes in human DNA, provided the data for mapping the three billion chemical base pairs that form genetic makeup of the human species. This article questions, how can we begin to understand the ethical implications of the development of a science which can alter the very fabric of human lives in the context of allocation of resources to the beneficiaries of genetic engineering? Subsequently, the *Genetic Information Non-discrimination Act* of 2008 was passed in the United States which sought to eliminate unfair discrimination based on genetic information [12]. The issue of genetic determination is inherent in the development of gene therapies and the use of genetic data in a consumer-driven society. It is common knowledge that environment does modify the expression of genetic information, however, to what extent is it feasible to make medical-ethical or social-policy decisions on the basis of genetic information and how is this informed by the ongoing irresolvable conundrum of 'nature or nature' debate? Is it the case as Busnelli (2002) suggests, that the question may become that of whether the 'mystery' of procreation is a value to be preserved or defended from the 'fanatics of artifice' [13]. Whilst the human genome project has produced undoubted medical benefits such as making possible new gene therapies, much of the research into the performance of specific human proteins in the DNA makeup is being carried-out by private companies. Such private development of gene therapies furthers the potential of the human species based on the understanding and manipulation of its genetic makeup but it also potentially risks the associated inequalities stemming from the uneven distribution of and access to medical technologies based on the consumer-driven model. Consequently, several further questions arise: what ethical models work best for the application of legislation following the sequencing of the human genome? Should law-makers and government policy in law-making by guided by the desired market-place outcomes balanced with issues of equity or human rights when making legislation controlling the use of genetic information or should they rather be influenced by what is known about the influence of 'basic' scientific discovery and debate?

The Human Genome project was completed in 2003 (ninety-seven per cent had been announced as read as a 'rough draft' by US President Bill Clinton and U.K. Prime Minister Tony Blair in 2000), as Winston states, "[t]he publication is a list of three billion letters that represent the chemical rungs of the DNA double helix, denoted by the letters [A (Adenine) which bonds with T (Thymine) and C (cytosine) which bonds with G (Guanine)]. It runs to over six hundred thousand pages of A4, and in book form would take up 270 feet of shelving space" [14]. Individual letters on the DNA helix differ between people only in the ratio of about one in a thousand, hence it is a difficult task to identify these within the code and to link them to the phenotype effect [15]. Furthermore, as Dawkins suggests the replication of DNA is so accurate that only about one-per cent in five million replication generations is miscopied [16]. Clearly DNA replication is an extremely precise and ordered process. Arguably full understanding of it is necessary before human intervention in the process of genetic re-creation is either possible or desirable. However, it is unlikely that gene therapy research will proceed in this way, rather it will be more like piecing together the pieces of a four dimensional puzzle, with a break-through in one area of gene sequencing impacting on another, all of which are understood as interacting in a complex manner on the human body through time. Thus the basis of medical research is in determining which malfunctioning gene causes a particular condition and designing ways of reversing it. The sequencing of Chromosome Twenty by the British 'Wellcome Trust Sanger Institute' has led to the search for genes involved in diabetes and leukaemia, for example.

Whilst the sequencing of the genome is necessary in understanding the genetic composition of human beings it is not as yet a finite achievement. There are a myriad of unexplored interactions of DNA and the expression of the phenotype. It is thought that researchers may work on this aspect of genetic research indefinitely despite the use of increasingly powerful computing capabilities. Furthermore, mapping the sequence of the human genome leaves open the question of the potential for the development of gene therapies to control and cure disease and of genetic procreation (creation by genetic design). How then are these dimensions to be considered, or protected by ethics and law in public policy if their potential is not yet known? Furthermore, to what extent can and should genetic determinism influence public policy and law, after all, knowing the basis of the genome sequence is one very significant and complex scientific achievement but understanding gene interaction and phenotype expression another, given the relatively unquantifiable and potentially unique factors of human interaction with environment. As de Melo-Martin (2005) has suggested, ". . . arguing against genetic determinism is like battling the undead: no matter how many times you stab them, they keep coming at you. This is the case despite practically every geneticist alive having disavowed genetic determinism" [17]. The relationship between genes and gene expression (nature or nurture) is complex and non-linear as genes interact not only with each other but also famously in response to changes in the internal and external human environment through processes of natural selection and evolution. It is this area, of mapping the genome and the complex interactions of gene expression that in medical research promises to benefit future generations in the fighting of disease. The focus of the next chapter is in describing in broad scope how the psychologists might view the influences of evolution on the development of the human mind.

Chapter 2

FROM BRAWN TO BRAINS

The human race used to be stronger. In evolutionary terms we are becoming taller and weaker as a species. We have grown larger brains to replace our ancestors' brawn. But at what rate are our brains developing in evolutionary terms? If we begin to look at the human species in terms of evolutionary psychology, what perspectives does it afford us on our species' development?

Evolutionary psychology is a theory of nature which claims that human psychological characteristics are adaptations from the Pleistocene[5] environment from which people evolved. Before the human brain became larger and we ascended to the top of the food chain, our ancestors were somewhere in the middle of the animal kingdom. Before we developed tool and language use and learned to control fire, and to cook food, we could not compete with the natural prowess of the top predators. As Harare (2014) states,

> For millions of years, humans hunted smaller creatures and gathered what they could, all the while being hunted by larger predators. It was only 400,000 years ago that several species of man began to hunt large game on a regular basis, and only the last 100,000 years - with the rise of Homo Sapiens - that man jumped to the top of the food chain. By shortening the intestines and decreasing their energy consumption,

---

[5] The Pleistocene epoch lasted for approximately two-and-a-half million years, a period of repeated glaciations, until approximately twelve thousand years ago. It was a term introduced in 1838 by the geologist Charles Lyell meaning 'most new.'

cooking inadvertently opened the way to the jumbo brains of the Neanderthals and Sapiens [18].

But what caused these evolutionary 'leaps'? An incremental process of behavioural (and hence also cultural) adaptation, or a sudden mutation in response to environmental stress? We don't know for sure. The tenets of evolutionary theory are congruent with Darwin's theory of natural selection which holds that for any given population there is variability in traits, there is differential reproduction, and there is heredity, all three of which lead to evolution by natural selection. (Other mechanisms of evolution include mutation, migration and genetic drift). For example, imagine the following evolutionary scenario. Amongst a social population, some people experience physiological phenomena that make them more grounded in themselves in response to stress, others don't. Those that do have a slight advantage in coping with new situations that predisposes them to be more positive in outlook and sociable, making it easier to find friends and partners. The children of the phenomenologically 'grounded' individuals pass on this trait to their children, hence there is a gradual selection for psychological 'groundedness' amongst a population. A population of people experience their lives as being grounded and more sociable. It spreads as it is a slight advantage over being anxious and nervous. Evolutionary psychology is a grand theory of the human mind, that advocates that explanations for human behaviour and the development of mental faculties are made in context of their behavioural origins, and that even neural mechanisms may be proximal cues for behaviour based on conditional learning from evolutionary origins.

The human mind is one of the most complex systems on Earth (perhaps the mind is a sub-system of being human and perhaps being human is a sub-system of life in the Universe, and it may be that linking all the subsystems together produces a gigantic human meta-mind), and there are currently no more complex systems possessed by competing species at higher levels to humans that are known to science [19]. Although evolutionary psychologists currently favour explanatory approaches to the mind based on psychological mechanisms understood in terms of computational theory and a form of universal modularity among homo sapiens, one of the difficulties lies with reconciling this approach to biological systems. At the moment this involves the decoupling of the metaphors used to explain mental functioning from their biological substrates. Rather than think of a complex bio-chemical entwining, the mind is understood as a vast array of mental modules based on evolutionary functions, akin to that of a 'Swiss-Army Knife.' These include:

Firstly, adaptations arising from genetic selection, secondly; an inherited mechanism that constitutes a 'single panhuman design' stemming from millennia of hunter-gatherer subsistence; and thirdly, the psychological universals that constitute human nature [20]. As Mossman (2014) relates, "[t]he human brain has about 10^{14} synapses connecting 10^{11} neurons" [21]. Emergent properties such as intelligence and consciousness don't seem to be derived from understand individual cells or parts of the brain alone but are instead likely the product of neural network connectedness and component inter-actions in multi-networks acting in complex non-linear ways. Mossman (2014) observes that neural functioning encodes information by the following processes:

> First, oscillating resting membrane potential are overlaid by input voltages that determine whether an action potential is generated. Second, information is encoded in the variations in spatial and temporal frequencies of action potentials. Third, neural networks have plastic. The number and sensitivity of synapses change as physiological requirements or external environmental conditions dictate. Fourth, synaptic membrane potentials are graded as a consequence of excitatory and inhibitory neurotransmitters [22].

The emphasis of this approach is on a causal-explanatory role of the evolutionary origins of behaviour. However, evolutionary psychologists are reluctant to attribute changes to the evolutionary psychological mechanism over the last ten thousand years of the Holocene period, arguing that although genetic adaptation can have been seen during this time (such as tolerance for Lactose), there has been little functional change to brain mechanisms or brain 'architecture' itself during this period. Thus evolutionary psychology is seen as a "convenient framework for discussing adaptiveness or as a way to integrate diverse ideas, rather than as a paradigm for developing testable behaviour" [23]. Adaptive changes to brain architecture are distinct (although ultimately related to) concepts of neuroplasticity.

Nevertheless, the human brain does possess some physiological adaptivity at the level of synaptic functioning. For example, neurons associated with ocular sensitivity can suppressed or activated by use. If a neonate individual does not use their sensory functioning at a critical time, then the neural circuitry associated with its use will not become activated. Similarly, people who have lost brain function in some areas have recovered the same function by using other areas of cortex. But these examples are distinct from the modular and architectural form of the brain and its components as the result of a through time, diachronic process of evolutionary adaptation. Thus the brain *does* adapt but it is doubtful whether the consequences of agricultural society and modernisation have wrought adaptive changes on the brain architecture in the relatively short time that they have been the overriding environmental conditions for the many peoples of the world. As Diamond (2012) states, "[t]raditional lifestyles are what shaped us and caused us to be what we are now. The shift from hunter-gathering to farming began only about 11,000 years ago; the first metal tools were produced only around 5400 years ago. 'Modern' conditions have prevailed, even just locally, for only a tiny fraction of human history; all human societies have been traditional for far longer than any society has been modern" [24]. Hence we might expect to find minor physiological adaptivity within the ten thousand years of the Holocene and patchily, not across the entire spectrum of humanity in a uniform way. Genetic adaptation operates in populations among 'fields of influence' as people interact and form families.

Table of Inferences for Evolutionary Psychology [25]

All species are capable of overproducing offspring.
The size of populations of individuals tends to remain relatively stable over time.
Resources for supporting individuals are limited.
Individuals differ on traits (i.e., adaptations) that enable them to survive and reproduce.
At least some of the variation in these traits is heritable.
Pre-existing adaptations, sometimes called preadaptation's, provide stepping stones and limits to the solutions natural selection can provide.
Genetic Preprograming of evolved traits - ontogeny of a trait is influenced by environmental factors, biologists refer to it as a facultative trait and label the genes involved in the development of facultative genes.

Evolutionary psychology studies the through-time modular development of the human brain. In part a deductive science, it attempts to explain how the human mind came to work as it does, to take the functional form it has, as a product of natural selection. The ability to describe the human mind and behaviour as a product of natural across-time genetic trait development is an attractive one, because of the promise of explaining how for the human brain, 'form follows function.' In the genetic programming of evolved traits, the architecture of the human mind is conceived of as being a layered system with basic survival functions regulating biological functioning, superseded by higher order abilities such as emotions and feelings and the cognitive symbolic manipulation that takes place in the synaptic neuro-chemistry of the neo-cortex. As Wilke and Mata (2012) state, "[e]volutionary rationality holds . . . that it is important to consider the match between the mind and the past environments in which the mind evolved" [26]. If we can infer how the human brain has adapted to its current form, we may be able to intuit how it may adapt and develop in the future. But the fact remains that although emerging technologies may shape our subjectivities they do not cause changes in our brain architecture.

Thus it is the study of conditions of ancestral populations that rendered some behaviour adaptive and others not adaptive and that mechanisms of natural selection are shaped to produce adaptive behaviours; and that processes of evolution shaped the way that human's functions now [27]. Given the survival and adaptation of homo sapiens from the many branches of proto-humanoids over the past four million or so years, it is possible that some human features have become less complex while others have become more complex, this is because evolution does not always proceed from the simple to the complex, rather "evolution often results in loss or reduced functionality of structures, including eyes" [28]. It possible that some now lost branches of humanoids had subtly different subjectivities of sensory perception and that even some forms of sensory experience have become more acute to humans in present adaptations and vice versa – the human olfactory bulb is larger than that of the Neanderthals implying that human sense of smell could be stronger [29]. Two common problems with evolutionary psychology is the length of time it takes to achieve (and thereby through study recognise) genetic adaptation of the brain architecture, which is generally thought to be an occurrence of human development during the Pleistocene era; and also the apparent contradiction between the logic of 'natural selection' and the preference for certain human behaviours such as altruism and that would seem to defy it. But as Pinker (1999) observes, "[t]hree hundred thousand generations and up to ten megabytes of potential genetic information are enough to revamp a mind considerably" [30]. Indeed four hundred generations and ten thousand years may be sufficient for some forms of genetic adaptation to take place and spread in a sub-population. The second issue, 'sacrifice' is understood as being any action such as food care or the protection of group members which results in the reduction of the reproductive success of the performer but conversely increases the reproductive success of the recipient.

The Lorenz (1966) thesis is that individuals adapt and evolve for the benefit of the group. However, there are other kinds of 'group centred' theories of evolution that explain how it is that the process of natural selection acts at a group level and is distributed through a population in a general sense rather than simply at the level of unique competition between individuals. Although in practice some pass on their genes and some don't, those that don't may not have any significant difference *at the adaptive level* from those that do. Chance, random event, and in human populations, social construction deriving from personality and behavioural expression that is not genetically based all play a part. The operation of natural selection on and between groups rather than individuals was also affirmed by Wynne-Edwards (1962) [31]. Related theories include, Hamilton's 'kinship selection theory'[6] (Hamilton 1964), parental investment theory[7], 'sexual-selection theory'[8] (Darwin 1871/1898) [32], and Trivers' (1971) 'reciprocity theory'[9] [33]. These theories are all versions of the 'fitness for the group 'interpretation of inheritable traits for natural selection. These concepts are particularly observable in the animal kingdom but the difference between the altruistic behaviour of bees who serve the colony and that of people is the complexity of the human brain and the fact that human personality and behaviour *overrides genetic determinism.*

The standard view of natural evolution holds that in intrasexual selection, competition between members of one sex (males) compete for access to the other sex (females) and that this produces selection pressure for increased physical size, organs of threat and aggressiveness in members of the competing sex. And indeed men are on average seven per cent larger than women and do not live as long [34]. The limits of genetic diversity are defined by a rule of thumb which states that fifty breeding individuals are needed to

---

[6] Kin selection was proposed by Darwin (1859) in *The Origin of the Species* as a concept whereby selection benefit to related organisms allow a trait to evolve but not an individual (for example honey bee drones) J.B.S. Haldane explained the mathematics of it while W. D. Hamilton popularised in in 1964. Hamilton's rule proposes that genes will increase in frequency by kin selection, when the genetic relatedness between the receiver of a benefit and the provider of it multiplied by the benefit to the receiver are greater than the reproductive cots of the provider of the benefit. It is observable in animal behaviour but human personality is more complex.

[7] Parental investment is a form of sexual selection in which parental expenditure benefits an offspring to a cost to other areas of fitness.

[8] Sexual selection has two forms. Members of one biological sex chose members of the other sex to reproduce with and compete with members of the same sex for access to members to the other sex. Some members have better success at reproducing than others.

[9] Reciprocity holds that people repay in kind what another person has given them. This applies to the obligation of favours and gifts, for example. It also applies to rewarding kind actions and not unkind actions.

avoid inbreeding depression, and five-hundred are needed in order for a population to adapt, though five hundred to five thousand is the preferred ratio for evolutionary adaptation to take place allowing for random losses as genes are passed on between generations (BBC, 2016). In trait selection, called epigraphic selection, members of one sex generate competition between members of the other sex by expressing choice of a mating partner. On the other hand, although organisms which have evolved in a population have done so because their adaptations provided ancestral organisms with a greater fitness, it is nevertheless false to say that these organisms evolved to maximise inclusive fitness. They responded to the environment to maximise the chance of perpetuating the genome. There is also a distinction between 'innate and operational' adaptations and a correlation between behavioural differences with an organism's fitness (and reproductive success) in an environment in which a trait evolved [35].

It is not so much that humans are evolving toward a greater perfection (much as though we might hope to think so) but rather we are evolving in such a way as to give the genome the best chance of replicating in a world governed by physical laws. It will also happen to be that evolution will not evolve a human population that did not also have a capacity for self-preservation, which in its myriad forms might be what we call a moral imperative. Biology also has a fragile robustness with human behaviour with complex gradations between what is useful, tolerable and undesirable from an evolutionary perspective. Pseudo-pathologies, for example, such as traits of excessive jealousy and anorexic behaviour in women, which have an effect on human behaviour and at least part of the human lifespan, probably have minimal inheritability. However, quasi-normal behaviours such as short birth intervals (the intervals between having children) and large families which acted so as to detract from ancestral fitness are more easily tolerated in the contemporary human environment. However, true pathologies are deleterious. Plenty of people are individualistic in society but sociopaths are literally hard to live with. All of which perhaps means that we are becoming more sociable and communicative peoples. Certainly our current technological adoption of social media and the inter-connectivity of the worldwide web would seem to support the latter view [36].

As Kendrick (2002) states, natural selection has 'sculpted' not only our physical bodies as living organisms but also cognitive programs to operate them. Evolutionary psychologists have argued that the human brain is based on modular mechanisms [37]. For example, human beings have evolved specialised organs designed to process sound waves, light waves, odours, taste, temperature as stimuli from the environment and the brain is a highly developed mechanism for analysing visual input, colour, shape, movement and depth.

Pinker (1999) has argued that the criteria for a brain module include factors such as: Firstly, functional engineering, what it needs to perform normal functions; secondly, biological anthropology -- what people need to do in everyday lives; and thirdly, developmental psychology to solve neurological problems, all of which occur within an evolved system of neural architecture [38].

Goldfinch (2015) considers biological altruism as being a behaviour that benefits someone else at a cost to the self but it poses a problem for natural selection in arguing how it is preserved [39]. As we have discussed it is 'preserved' because it is a distributed tendency that emerges from group behaviours. Wynne-Edwards (1962) [40] argues that it is a kinship group weak evolutionary force; and Trivers (1971) that it pays an organism to help another [41]. Not only are humans social people but we learn that cooperation as well as competition is good for us in general. But when and how did we get like this? The current general consensus is that modern homo sapiens evolved between two-hundred thousand and seventy-thousand years ago, which is the blink of an eye in evolutionary terms.

Physiological development of the human brain occurred in the time depth of the Pleistocene era. Aside for physiological adaptations at the substrate level, the Holocene era of the last ten thousand years (approximately equivalent to four hundred generations) is, as we have encountered thought to be too short a time for large scale genetic architectural changes to occur. So, on the one-hand as Goldfinch (2015) argues, the "[p]ossibility of post-Pleistocene selection for psychological adaptations, over the 10000 years of the Holocene, [is unlikely]. We have spent the last around over 10000 in the era of agriculture Holocene and industrialisation in only the last 200" [42]. But if change did occur in response to modern society and its relative peace and affordance, what change might that be?

The 'flight, freeze or fight' amygdala anxiety response that has suited human behaviour in ninety-nine per cent of its time in hunter-gatherer societies as a response to ecological threats, dangerous animals, disasters, floods, storms, landslides fires, droughts, famines could be unsuited to the modern era in which many of these ecological concerns have become more tolerable. Certain features of the modern industrial and post-industrial environment are novel, these include contraception, international travel and eating habits (current predominance of fatty and sugary foods); similarly those threats associated with climate change, competition for natural resource (water, food), ozone depletion, and increased ultra-violet intensity have largely yet to result in trait selection over so short a time scale despite the fact that theories of genetics and experiments on other organisms show that adaptations of traits can occur in spans of less than one hundred generations [43].

However, contrary to Goldfinch, Merkel-Bobrov, Gilbert, Evans, Vallender, Anderson, Hudson, Tishkoff, and Lahn (2005) revealed that a "variant of the gene ASPM regulator of brain size" arose only around eight-to-five hundred years ago and is now carried by a quarter of the world's population, hence arguably being evidence for the brains rapid evolution [44]. Certainly Holocene specific adaptations such as lactose tolerance as a result of the increase in dairy products since the agricultural revolution and the domestication of cattle have been observed by Burger, Kirchner, Bramanti, Haak, and Thomas, (2007) [45] despite the fact that lactose intolerance is more common, particularly in South-East Asia [46].

Hence there is no reason to suggest that modification of cognitive adaptations might not also occur (for example, suppression of amygdala functioning) as a response to altered environmental conditions. Furthermore, if less extremely primed 'fight or flight' mechanisms are called for then relational and language abilities might offer some selective advantage for those who are better able to negotiate, communicate and understand and empathise with others, providing environment conditions remain less extreme. As Pinker (1999) observes, like Hararer, perhaps somewhat pessimistically, for our celebrated cerebral dominance over the animal kingdom and perhaps the ecosystem, "[t]he modern mind is adapted to the Stone-Age, not the computer age . . . [people] are not wired to cope with anonymous crowds, schooling, written language, government, police, courts, armies, modern medicine, formal social institutions, high technology, and other newcomers to the human experience" [47]. Yet our language use is becoming both more homogenous and varied. We communicate more not less as a species among and between sub-populations, cities, nations and between countries. Indeed, there are some forms of mass and meta-communication that are very recent products of information and computer technology, the internet and globalisation – such as *Facebook,* and *Twitter.* Indeed, as Benitez-Burraco and Boeckx (2015) suggest, "[c]ertainly, the number of functional links important for language evolution is becoming bigger as time goes by" [48]. Adaptiveness for memory and language modules might become heritable traits that facilitate reproductive survival over aggressiveness and or people prone to extreme sensitivities of anxiety.

According to Hampton (2009) the Environment of evolutionary adaptiveness (EEA) is a time period one million eight-hundred-thousand years since the emergence of genus *Homo sapiens* and the human mind complex system of computational mechanisms selected for by natural selection [49]. This period is more than long enough to entail the morphological variability (and possible interactions) between hominid taxa. Homo sapiens and Neanderthals for example may have co-existed for approximately one hundred and fifty thousand years before Neanderthals were selected out some fifty thousand years ago. Despite the fact that modern humans may share up to four per cent of Neanderthal genes, evolutionary biologists are unsure whether Neanderthals and humans shared biological compatibility, minor genetic differences affecting the physiology of cellular immunities may have prevented human females from carrying a foetus from a Neanderthal to full term [50]. As Sankaraman, Patterson, Li, Pääbo and Reich (2012) relate, "[e]vidence, both for and against interbreeding, have been put forth based on the analysis of modern human DNA. Although mitochondrial DNA from multiple Neanderthals has shown that Neanderthals fall outside the range of modern human variation, low levels of gene flow cannot be excluded" [51]. If interbreeding did occur, then "[i]n that sense, Neanderthals did not go extinct, even though their distinctive morphology did disappear" [52]. Physiological cognitive differences between Neanderthals and humans may not have been great but a possible reason for the isolation of the Neanderthal group, was the interaction between "biological and behavioural cold adaptations, in context of extreme climatic fluctuations in the Upper Pleistocene" [53]. Sapiens responses to climate variations were more adaptive due to their ability with technologies for manufacturing clothing [54].

Human and Neanderthal differences cannot be explained only from archaeological markers such as bone technologies and blade-based lithics, or indeed only on the basis of more ambiguous markers of cognitive or linguistic abilities (with low archaeological evidence). Instead, Gilligan (2007) argues that humans become more adaptive to extremes of weather variation because of clothing technology, which allowed them to withstand cold stress to a greater degree in Ice age Europe. This corresponds to the greater precision of lithic (blade) technology amongst humans in this period. Humans also had greater control of fire, and use of sheltered sites, more specialised hunting abilities, made personal adornments, used pigments, used grindstones, developed parietal art, had wider geographic variation, and had increased artefact diversity and could mine [55]. The use of clothing technologies also displaced images, argues Gilligan (2007), from the body, onto garments and external surroundings, becoming more visible in archaeological trace [56].

Behavioural difference as the product of evolved psychological mechanisms depends on internal and environmental input for development, activation and expression as Darwinian traits can be inherited by offspring from parents, as traits which aided an organism's survival and reproduction were transmitted to future generations at greater uptake frequencies [57]. Genetic adaptations can occur in three ways. Adaptations and inherited characteristics that reliably solved problems are related to survival and reproduction better than competing alternatives, during the same time period; secondly by-products, and artefacts without functional value persist because they are coupled with adaptations, (for example, for many people in the world, the fear of harmless snakes); and thirdly, noise, and variations in a given characteristic due to random environmental events or genetic mutations (for example random low base rate fears) [58]. Some adaptations like the relatively short human intestine for digesting cooked food arose from humans solving problems of controlling fire, food selection and avoiding substances containing toxins [59].

The retrospective cultural pattern of evolution, sees that the human body may change due to functional selection as humans interact with the environment and between themselves, and it raises such haunting questions as whether human memory is sensitive to content relevant to evolution [60]. There is some informed speculation that in traditional societies some cultural 'stories' have been communicated across ten thousand years of time and down through about four-hundred generations, in Australian aboriginal society for example, which is coincidental also with the periodicity of evolutionary adaptation. The human mind is a complex integrated assembly of many functionally specialised psychological adaptations that evolved in response to problems of parental investment, kinship, co-operation, selective aggression and negotiating status. Put simply, the mind and brain through body preserves and passes on the genome but it also the product of a group process and enables our sociality.

Chapter 3

A MINDSCAPE

One of the best tricks our ancestors taught us was how to make tools and use fire, so we no-longer had to fend away the dangerous world at spear point. It also enabled us to cook food and resulted in an adaptation for a shorter human intestine. Our ancestors enabled us to eat better and to adapt in evolutionary terms to think and reflect. The growth of the human neo-cortex has allowed us unprecedented control over the environment, which we can shape in ways we think best fit as a society. But reflection allows us to use some of this knowledge to think about the environment differently than just as something outside of us and separate from us. In this respect we can use some of the best aspects of traditional society (an attitude towards knowing how the environment might sustain us, and how we can read it, in relation to ourselves) while throwing away the worst – the increased aggression and religious mumbo-jumbo. Much as our ancestors arguably always have, we might be able to harness technology and ways of seeing to blend the traditional and the modern. As such environmental communication is concerned with how people create their own language and signs through symbol, sound, or non-verbal body language and how they perceive, encounter, interpret and use the signs created by others within their ecological niche, and how they might adapt their environment to ensure their preservation, sustenance and well-being.

Surrounding the human being is a semiosphere of information from the environment and with which people interact through senses and perceptions. People, organizations, societies and cultures construct, receive, interpret, understand and use messages about and within the environment involving human interactions, thoughts and informational exchanges, and people negotiate their living space through relationships, constructions and activities which sustain and support them. As Dillman (n.d) suggests, "[n]o act of communication occurs independently of its environment" [61]. As such the semiosphere of interaction of people with the environment, in which information is received and exchanged is complex and interdisciplinary.

Varied Characteristics of Environmental Communication

There are many ways in which biotic organisms – humans, animals, and vegetation – express themselves in their environment, and exchange and communicate information in order to fulfil their niche biological functions as part of a distributed process of continuously evolving organic life-form inhabiting ecosystems on planet earth. This may take the form of adaptation to environmental conditions (such as noise) or presence and absence of light; the complex permutations of perception and cognition in which the environments shapes language and thought (and vice versa); the development of haptic (touch) communication in both real and simulated contexts; the impact of fluctuations in physical properties of ecosystems all effect our response to and the gradual selection for sensory features. The light band spectrum, the noise decibel distribution; the effects of weather formation and/or the effects of technology (ergonomics) on human functioning and lifestyle, all constant background permutations of our environmental negotiation.

Because we are social, one of the basic features of human adaptivity is 'face recognition.' As Pinker (1999) observes: "[f]aces are so important that natural selection gave us a processor that registers the kinds of geometric contours and ratios needed to tell them apart. Babies lock onto face like patterns, but not onto other complex and symmetrical arrangements, when they are only thirty minutes old, and quickly learn to recognise their mothers, perhaps as early as the second day of life" [62]. Hence face recognition is intrinsically linked to the sociality of our species. It seems to be innate, and a function of our environmental behaviours that differentiates us.

However, neotony aside, in order to survive animals must be able to detect/assess/send signals in their environment, to negotiate affordances and meet needs. This results in complex patterns of experiential expression, some of which is intra and inter species communication. Consequently, animals evolve signals that distinguish them in effective feedback loops in the environments they are used in. Ord et al., (2011) studied the effect of environmental noise on the communication of *amolis* lizards from radiations on Porta Rico and Jamaica surmised that ". . . [t]hose species communicating to distant receivers and in habitats in which light levels are low (e.g., in full shade) should be under considerable selection pressure to time signals to exploit momentary windows in noise [63]. If so, this might result in predictable variation in signal timing among species living in different habitats and with different ecologies, irrespective of island origin [64]. Consequently, the same adaptive patterns may be observable in the same conditions amongst distantly related species, such as communicative responses and camouflage patterns for example.

Relatedly, zoologists have observed the ability of some animals, for example birds and mammals to navigate by using the Earth's magnetic fields. There are three main mechanisms proposed for this: Electromagnetic induction, ferrimagnetism, and chemical reactions involving pairs of radicals [65]. Magneto-reception has been attributed, in the small and cheerful flighted bird, the Robin (whose Latin name is *Erithacus rubecula* in Europe and *Turdus Migratorius* in America), to the presence of the molecule cryptochrome [66] within their eyes, which when stimulated by magnetic fields can provide an information overlay that makes some parts of the bird's vision perceived as being brighter, hence easier to fly towards. Cryptochrome may not be the only magnetic sensitive molecule, there is some evidence that Pigeons may have neurons which are magnetic-field sensitive. Hence magneto-reception (the ability to perceive magnetic fields) may be more broadly spread in the animal kingdom. Do humans retain any of this ability to perceive or sense magnetic fields? Would it provide an advantage for evolutionary selection? Possibly, if it increased the chances of finding 'home' when lost, or in navigating without technological aid. Some people do seem to be better at orientating towards the points of the compass than others but arguably this is simple awareness and a learned behaviour, rather than an innate disposition.

Human beings are reasonable malleable physically despite in the absence of exo-skeletons being relatively fragile. As Harare (2014) states, "[h]umans emerge from the womb like molten glass from a furnace. They can be spun, stretched and shaped with a surprising degree of freedom" [67]. Providing they receive vital years of care and nurturing, human beings are reasonably robust and versatile. As Broditsky (2011) suggests, a predominating feature of human intelligence is adaptability which is the capacity to invent and rearrange conceptions to suit the changing demands of the environment [68]. Languages reflect both this capacity and the diversity of its effects. In this way language molds the individual into the environment. Broditsky (2011) regards language as an “invaluable guidebook developed and honed by our ancestors” [69]. Differences in the environment create differences in language and thinking and vice versa. Human technology use may reflect the two characteristics of communication: Firstly, that it can be deterministic in nature – reinforcing social and structural norms; and secondly that it is emergent and capable of creating new meanings [70].

One recent adaptation to virtual technologies is the incorporation of haptic ‘touch’ feedback devices. These are used for example in medical devices for minimal invasive surgery, remote manipulation by robots in space, and also in collaborative virtual environments (CVEs). These are digital spaces that allow remote users to work together and to experience tactile communication exchanges in doing so. However, collaboration also requires a shared mental representation. This is referred to as a frame of reference or common ground. As Chellali et al., (2011) argue, haptic feedback is a direct form of human interaction – it can be used to express feelings of closeness, intimacy and trust [71]. So the basic communication medium of touch, carries information, it does work, but it also carries intentionality, empathy and emotion.

According to Rasmussen (1983) there are three categories of human behaviours – skills, rules, knowledge. The sensory-motor performances are based at the skills level – shared by a reflex behaviour, haptic communication requires physical contact in this modality and is possible even in remote communication but may involve the oldest regions of the brain. Brier (2013) proposes a field of ‘cybersemiotics’ in which ‘semiotic cognition and communication’ forms a ‘basic sort of reality in which all . . . knowledge is created’ and then suggests that knowledge develops from four kinds of human reality: Firstly, a surrounding nature described by the physical and chemical sciences; human corporeality described by life sciences such as biology and medicine; an inner world of subjective experience described by phenomenologically based investigations of the social world described by social science [72]. At the basis of all these ways of perceiving and dividing the human semiosphere is the complexity of information that informs the human ontology.

Environmental Communication and the Concept of Place

The experience of place in environmental communication is a dimension formed by peoples' relationship with physical settings and activities within the semiosphere of symbols and signs, images and perceptions, thoughts and emotions that obtain to the particularities of place. As Najafi and Shariff (2011) point out, 'place identity' and 'sense of place' are sometimes used synonymously and describe the quality of peoples' place relationships, how they relate to their habitats, psychologically and physically. However, the two terms differ in subtle ways. Firstly, human-place bonding refers to the attachment with or relationship to a specific semiosphere of locale, while sense of place may refer to the assignation of ties, perceptions, experiences in a relation to a place in terms of an overarching impression [73]. Sense of place in environmental communication is important to the quality of the environment and human integrity.

Furthermore, place can be distinguished from space in as much as it represents an affective bond between a person and a particular setting. The concept of 'place' is imbued with human values whereas 'space' may be a natural term, however, places may be in space, and have a unique character. Consequently 'sense of place' helps to maintain the quality of the environment, it may be involved in connecting to people to shared experience and it may help to consolidate community ties. However, sometimes places have a lack of identity or no 'placedness' and convey the minimum of feelings and meanings to people and may be indistinguishable from any other similar environment, in such space one can usually still discern basic features such as perspective but no implicit feeling of relatedness or relatability. However, designers, architects and landscapers, may create a sense of place where none existed before, some spaces featuring 'placedness' occur naturally however. The added element which changes environment to place is the 'experience' provided by the phenomenology of perception, within a semiosphere of signs and symbols. This being in the world is related to placedness a fundamental quality of psychological existence and may be a product of experiences of quality in existential space. Najafi and Shariff (2011) explain how the phenomenologists concept of 'Topophilia' is used to represent 'character of place' and 'spirit of place' [74]. Topophilia is the experience of place may have several levels – the first is familiarity which may or may not be associated with depth of feeling and meaning. The second level is ordinary familiarity, perhaps the unconscious perception of place, experienced more on a collective and cultural level than a personal level and signifying a 'deep' attachment to place and close attention to its symbols. The third level is that of 'profound familiarity' involving the 'existential insideness' of a person and integration with place [75].

The communication of relatedness to place and sense of place is important to environmental psychologists as the physical environment plays a role in both the immediate and long term impact of human behaviour and mental health. Consequently, the design of place involves both emotional and functional qualities. Thus 'sense of place' encompasses the relation between humans and place. Najafi and Shariff (2011) suggest it is defined by three elements – location, landscape and personal involvement and furthermore, Rogen et al., (2005) define a psychologically comfortable environment as being – legible, capable of qualitative perception and preferable – compatible for human living [76].

Najafi and Sharrif (2008) identifies seven levels of human interaction with 'sense of place' in the environment [77]:

- Evolution in an ecosystem
- Knowledge of being located in a place
- Belonging to a place
- Attachment to a place
- Identifying with a place and its purposes
- Involvement in a place
- Sacrifice for a place
- Not having any sense of place

Involvement with sense of place as a phenomenological experience is thus both psychological and social/interactional and physical. Peoples' relationship with place may also be transactional – people either give or take or experience positive or negative associations with place. There may be several factors influencing place attachment:

1. The emotion of feeling people have for place,
2. Socio-demographic characteristics,
3. Environmental experiences,
4. The nature and type of peoples' involvement with place,
5. Peoples' degree of familiarity with place,
6. Peoples' expertise with place,
7. The experience of culture in relation to place,
8. Peoples' satisfaction with place,

9. Peoples' preference or attachment to place,
10. The kinds of activities people experience with place,
11. the phenomenological experience of place itself,
12. The extent to which place supports or inspires concepts of self,
13. The extent to which an environment is sustaining,
14. The extent to which an environment is changing and/or supports successful adaptation.

ENVIRONMENTAL COMMUNICATION AND THE HUMAN ORGANISM

Environmental communication spans both the material and non-material in human interaction, behavior and thought, the physical and the psychological and even the spiritual – the conscious experience of growth. As Simpson (1964) suggests, "living things have been affected for . . . billions of years by historical processes . . . results for those processes are systems different in kind from any nonliving systems and almost incomparably more complicated" [78]. While these process are historical they are also evolutionary – a part of a natural order, that emerges from the complexity of life. This suggests that the laws of perception, behavior and communication are ecological and defined in terms of the organism and its environment – a semiosphere [79]. As Stewart (2007) suggests, "[i]ndividual psychologists believe that . . . lifestyle depends upon myriad organismic and environmental influences that people experience as they pursue the tasks of family, work and community" [80]. Adler (1938) for example maintained that organismic and environmental variables are inherently interactive and are significant in affecting personality [81]. The relationship between a conscious organism and its function are complex – the actions of the organisms are intrinsically to do with its functioning and 'construction.' For German enlightenment philosopher Immanuel Kant, an organism can be distinguished from a machine for this reason, " . . . the parts act together to meet the thing's purpose; their actions, however, have nothing to do with the thing's construction. The thing and its parts rely on efficient causes arising from outside themselves for their origin and function" [82]. On the one hand the human organism is 'built' by a natural process on the other hand much of this 'building' is accomplished without conscious intent.

Consequently, an organism that is capable of self-organization and which holds an ecological property are synonymous. The parts are not easily separated from outside of the whole they comprise. This is a definition of complexity, such that the organism is not fractionable [83]. For example, take the self-organizing process in the evolution of a bird's wing. The propulsion and lift it offers the bird are not defined outside of the form and function of their whole [84]. As Francovich (2010) observes, the concept of 'agential realism,' "captures the paradox of the observer and the observed by means of the argument that relationship precedes related and that we as individuals/organisms are in a flux of continuous emergence that is fundamentally nondual" [85]. Whilst environmental communication may take a materialistic slant towards the investigation of the form and function of the human organism and the environment in which it is situated and which it co-creates. After all living organisms are subject to the same physical laws as any other entity. But how do we account for human cognition, language, the 'illusion' of free-will, consciousness, emotion, memory, imagination amongst other phenomenon of qualia? Through much of history this has led people to the believe that the conscious organism is divisible into theories of the non-living forces – however this deconstructivist tendency runs against the properties of an ecological system that it comprises the environment of which defines its purpose as supporting life and evolution. Living systems are generative, creative and non-linear also.

Petrusz and Turvey (2010) argue that there are four kinds of laws that define a 'science of the animate' in human psychological and environmental communication terms. These are: Firstly, human behavior is guided by the capacities of the organism-environment as a whole defined in terms of intentions or opportunities for realizing them. They may be understood in terms of lower level physical features and non-fractionable higher level features. Secondly, human behavior is guided at the ecological scale of the organism and environment not of their components. They are defined by a meaningful physics of semantic affordance. Thirdly, the human organism is more general than (not a special case of) the laws of an animate system. It is not the case that only laws that govern animate systems are those that govern their physical components. As Petrusz and Turvey (2010) suggest, "the causal entailment of animate systems exceeds that of physical (inanimate systems) [86]. Fourthly, they make reference to themselves in an impredicative manner, the fact that in human cognition and the semiosphere, perception and action can be conscious and self-referential as well as self-other referential.

From a perspective of environmental psychology, the study of human-environment relationships is a wide field and includes topics such as: territoriality, personal space, crowd effects, environmental stress, scholarly, business and work environments, home environments, environmental influences on behaviour, attachment to place, isolation and contained environments, the assessment of environments and also values beliefs and attitudes concerning the environment [87]. Thus the literature of environmental communication and psychology is informed by laboratory, virtual, field and ecological settings as was as from architecture, design and industry. There are three main paradigms. Firstly, the 'adjustment paradigm' which views the environment as physical, social, interpersonal, communicative influences on people as they adapt, respond and function in the environment, for example, to stress, arousal or stimulus [88]. The 'opportunity structures' paradigm studies how people function in a sociophysical environment to meet their needs, goals and wants and how they perform their roles within it. Thirdly, the sociocultural paradigm is the study of how culture and society shape the sociophysical environment and relationships within it [89]. Consequently, environment may shape behaviour and to some extent personality but it is also true that peoples' attitude and behaviour towards the environment can affect the psychological well-being of the individual and that of other people [90].

As Binnie-Dawson (1982) relates, the bio-social approach to environmental psychology and environmental stress is permissive of three main orientations towards the environment (after Wohlwill, 1970): Firstly, behaviour occurs within an environmental context which imposes condition and constraints on the functional of the individual within it; secondly, some environmental conditions such an over or under stimulation, climate extremes, will have more generalizable effects on system of response within the individual, and thirdly; behaviour is instigated by particular environmental attributes and characteristics [91].

Environmental adaptation is a social as well as physical process, social systems as well as physical systems lead to adaptation sin perceptual and cognitive abilities, attitudes and values which in-turn effect ecological survival [92]. As Lakoff (2011) suggests, people are nature – "nature is not separate from us . . . what is good is the use of nature that doesn't use up nature" [93]. We inhabit a biosphere which is a thin zone of atmosphere and marine and terrestrial ecology that sustains a process of evolution of organic life on Earth. Earth itself orbits a sun whose power and energy is derived from fusion, a process of proton-proton reaction in which hydrogen nuclei are converted to helium, which provides the energy for life on Earth through synthetic processes.

ENVIRONMENTAL INFLUENCES ON SOCIAL BEHAVIOUR

Humans respond to physical cues in their environment through perception by the 'guided' senses through nerve and neural systems from the semiosphere. As Stewart (2007) suggests, "'[b]ehavioural settings' refer to systems emerging from the temporal and spatial arrangement of social and physical characteristics of small-scale environments" [94]. One of the main environmental influences on behaviour is crowding. There is some evidence that crowding has a negative impact on the social behaviour of children. The reasons for this are that it may produce an adverse subjective experience on more vulnerable and younger children leading to relationship difficulties from restrictions on behavioural freedom, excessive stimulation, or intimacy that is too intense or inappropriate [95]. Under-population of the environment may also lead to less than optimum development of meaningful interpersonal roles. Under population theory holds that there will be a maximum amount of positive psychological benefits when there are more behavioural settings and roles to be fulfilled and experienced than there are individual people competing to assume those roles [96]. Under-populated social environments operate in a *centripetal* manner such that interpersonal communications and behaviours provide feedback, participation, cohesiveness to maintain a frame of social reference for any given task or activity [97]. Overpopulated social environments may lead to marginalisation through *centrifugal* social forces.

Gibson (1979) is attributed with founding the field of ecological psychology with influences on both cognitive psychology and environmental communication. Ecological psychologists characterise 'information' as the "patterning of energy that occurs as it passes through a medium and interacts with objects" (See Jordan, 2009) [98]. There is a quite profound difference between this model of human perception than that of cognitive psychologists. Instead of emphasising the ambiguity of information and the brain's capacity for specialisation in processing it, ecological psychologists distinguish less between the person and their environment, arguing instead that the brain and sensory system 'resonate to environmental information' [99]. Under this conceptualisation, the behavioural possibilities that a stimulus affords is equally as important as its perceptual qualities. Thus instead of emphasising the ability to perceive and cognate an object internally (to assess size, shape and colour) ecological psychologist emphasises the affordances of such perception in terms of the possibilities it offers the individual for use – what can be 'done' with an object.

As Jordan (2009) points out the ecological psychology view of human-object relations affordance is consistent with the theory of perception-action coupling termed 'Theory of Event' coding. Theory of event coding asserts that: Firstly, actions are planned in terms of the distal effects produced (i.e., outcome in the environment), and secondly, cognate perception, planning and action use overlapping neural resources [100]. This would imply that perception is inherently intentional because planning alters neural configurations of perception which in turn is consistent with the ecological psychology view that perception is made in terms of affordances (behavioural possibilities detected by perception [101]. Affordant properties are emergent contextual and base on perceptual perspectives, as Francovich (2010) suggests, " . . . the behaviour of any organism can be understood as melded or continuously fused to the medium, surfaces, and substances of contact experience . . . in continual emergence" [102].

Intentional induction, for example putting one's hand over one's eyes to scan the horizon is one such example. Ecological psychology also renders the explanation of experiencing the intentions, desires or beliefs of another in less fractal terms than computational approaches. We experience the mental contents of another person because we experience the distal effects they generate in the environmental and the actions and means that generate in the context of our own embodied action-effect contingencies. This differs from either the computational or cognitive models of mind in so much as, instead of being 'trapped' inside non-observable, symbol-manipulation systems, as is assumed in computational approaches, the resonance-based approach asserts intentionality as entailed in the continuous synergistic, multi-scale couplings that constitute intentional context including neutrally embodied action-effect regularities, neuro-muscular architectures, and the external regularities such architectures give rise to and are embedded within" [103]. This has large repercussions for environmental communication in so far as the ability to couple individual intentional contexts within group contexts is fundamental to an ability to generate, use and sustain sign systems [104].

As Craik (1977) suggests the perceptions an observer derives from a setting depend not only on physical attributes but other factors such as: Firstly, the cognitive set that is brought to bear on and is inspired from a space, secondly, how the place is encountered, and thirdly, available formats for responses, as well as affordances [105]. Furthermore, as Francovich (2010) relates it is inherently related to the creation of self and environment, "[s]elf [is] a dynamic emergent process that comprises the 'living present' . . . whereby reflective consciousness creates the symbolic world of spatial and temporal independence and thereby stabilizes the appearance of enduring subjects before static and/or dynamic objects (artefacts of reflective consciousness) in a timeless space regulated or metered by spaceless time" [106]. In this respect consciousness is like a 'platform' a station in the mind that is achieved initially by learning and conscious will and from thence becomes second place, like riding a bike. However, environmental perception is arguably at its most complex when the encounter is with other people. As Adams (2007) relates interrelating is fundamental to human culture and can be characterised in three ways: Firstly from the beginning of every presencing moment, secondly, from is forms the basis of human interaction – it is the progenitor of health, justice and compassion for humankind, thirdly, it is the 'ever-present' path by means of which we carry out our 'inter-existence,' fourthly, there is an ethical imperative to cultivate interrelation in a way which serves others and the non-human natural community [107]. Furthermore, arguably the socialisation of inter-relation contributed to the development of human consciousness deriving from environmental communication. As Francovich (2010) suggests:

> The significant gestures (Mead's term) that organisms employ are happening on an evolutionary ladder of increasing coordination via communication amongst entrained interlocutors that eventually, and through pressures of increased social complexity and evolved neurobiological structures, result in a 'phase shift' in communications. The conversation of gestures, according to Mead, results in homo sapiens internalizing via the significant symbol first specific to others (family) and then through a progressive process of role taking and role playing eventually come to the generalized other and creation of the reflective self. This internalised generalised other becomes the 'me' of the reflective consciousness and is the beginning of the deep patterning that we come to know as our subjective selves [108].

Increasingly it may be seen that human consciousness is 'learned.' We humans are born with certain innate propensities – for face recognition and generative grammar, for example, but it can be seen that consciousness is an acquired process that we are habituated towards. Consciousness is an experiential 'platform' that is a quality of experience achieved through socialisation and learned through habituation. We are 'taught' to become conscious by the attention of others. And if conscious is acquired and learned, what else can we humans learn? Are there innate properties of a socialised and inter-connected human meta-mind? These are not idle questions that may be addressed in the passage of time, the product of evolution in the not too distant human future. Consequently, the need to improve the ecological environment is not only a biological imperative but a social, technological and ethical one, requiring the exercise of authentic relationships. Inherently our use of language links us to the perceptual participation in the phenomenal world in a system of inter-relations signifying both the evolution of the human species and the individual development of the child and also in the moment of co-presence with others [109]. However, as Adams suggests, extreme versions of linguistic primacy may lead to the anthropogenic argument that humankind is both self-sufficient and self-sustaining [110]. Thus human communication and human culture are inter-related. They both have pattern, some form of order and consequently and evaluative dimension that is dynamic over time.

Chapter 4

CAMOUFLAGE

As with a biosphere which surrounds the planet, the zone or tangent in which life evolves, an organism also inhabits a semiosphere. The semiosphere forms the world of natural communication which stems from the organisms natural and biological features, consisting of the communications, sights, sounds, odours, movements, colours, electromagnetic fields, wave particle dualities, chemical signals and haptic touch perceptions [111]. The semiosphere is the domain of affordance. To describe a complexity of affordances, in this chapter we consider the concept and perceptual phenomenon of camouflage, which despite our body hair which protects us from some sunlight and cold (and which arguably we are adaptively loosing since the universal adoption of clothing) is appropriated by us from the natural world.

Camouflage is reasonably common in the animal kingdom, occuring among both predators and their prey, both in terrestrial and marine environments. Although octopi and squid (cephalopods) use camouflage through manipulation of light through colour-changing cells called chromatophores below the surface of the skin, the most famous exponent of camouflage in nature in the chameleon. The chromatophores contain elastic-like sacs of pigment which can be stretched and which respond to chemical signals from the nervous system and bloodstream. Cephalopods also have iridophores which are cellular stacks of reflecting plates that create iridescent colours and leucophores which can mirror back environmental colours. These animals' camouflage colourings are a form of affordance. This term was coined by psychologist J. J. Gibson (1977) and refers to the actionable properties between the world and an organism (usually a person or animal).

Affordances are a part of a natural relationship and do not need to be visible, or known. They can be discovered and can also be dangerous. Affordances describe among other things zones of continuity and discontinuity between organism and the environment. As Bertolotti and Magnani (2014) argue camouflage is part of an ecological problem which is about external agency, the notion of the continuity and discontinuity between an organism and their environment not being absolute, and the inter-relationship of organisms, and the fact that each organism is part of the other's ecology [112]. But how is camouflage defined? As Leonie (2010) states, ". . . camouflage can be defined as that modality of the invisible in which invisibility is not absolute, but relative to the relation between an invisible entity, an environment, and a viewer" [113]. Hence when an entity looks to an observer as entirely resembling their environment, camouflage gains its meaning valence. If the camouflage is entirely successful, the organism is not differentiated from the environment and effectively 'vanishes' in it. Camouflage thus inspires a form of perceptual illusion the perceiver.

Leonie (2010) argues that while camouflage is an invisible modality, it is also a modality of perfect resemblance to an environment, modelled from the invisibility of certain natural elements and some processes of weather. The affordances of camouflage may be costly against something and in protection or benefit of another entity [114]. It is an irony of the camouflaged organism or entity that it is both a subject and object and anti-subject and anti-object in its environment, not based on disappearance of entity but on opaqueness of it when presented to vision. [115]. Hence mist is a form of natural camouflage and at the high-technology end transformation optics has become adept at virtually extincting an object by redirecting the electromagnetic waves that impact it without disturbing surrounding environments. This acts so as to effectively cloak an object based on the concept of scattering cancellation to achieve electromagnetic transparency [116]. James Bond's Aston Martin had an effective cloaking device in the film *'Die Another Day,'* the camouflaging of which is now a not too distant proposition in reality.

As Webster (2015) asserts, camouflage is a prevalent feature of nature and has benefits for both prey and predators [117]. Both in its defensive and offensive forms camouflage has evolved as a protective shield to avoid visual perception and is understood as also referring to an ability to make an object, concept or even an idea to seem or appear different from what it is, or to make it disappear [118]. Camouflage thus is a concept that originates from biology to describe methods and strategies by which organisms dissimulate their presence in an environment, but it is a concept that is readily adopted in other domains and semantic fields as referring to concealment of one's position, meaning, intentions and or motivations [119]. Camouflage is used to conceal, disguise or counterfeit the presence of an object or thing in habitat to conceal revealing patterns and to signify false forms to disrupt enemy intelligence. Two states of camouflage are believed to be naturally occurring and ubiquitous in nature, these are broadly defined as 'background matching' and 'disruptive colouration' [120].

Background matching camouflage emerges as a property of the visual appearance of the environment and disruptive coloration is characterised by markings that break-up and distort information edges. Camouflage strategies may involve a combination of the two [121]. These characteristic traits of animal appearance in their environment provide concealment for auditory, chemical, tactile and visual sensoriums, operating as a selective advantage for both predators and prey to avoid detection (being amore a feature of the environment than those less successful) [122].

Camouflage enables entities to 'hide in plain sight.' Offensive camouflage allows predators to approach prey without being seen, whereas defensive camouflage allows prey to avoid being seen [123]. As Bertolotti et al., (2014) state, "[a]nimated beings do not merely 'contemplate' the environment uninterestedly but as survival machines they are actively involved in continuous problem solving activities" and not all features of the surrounding are equally relevant "when human beings look at the natural landscape, attention is automatically driven to search for and investigate little movements, glitters, and shadows that signify the presence of other life forms" [124]. So while humans are busily about the business of detecting the presence of others in the environment for conversation and trade and exchange (and rarely for fight or flight; hostility or empathy) they may also be about the business of 'fitting-in' of not standing out, of passing through a landscape or habitas causing little fuss and passing 'as if' unseen in the ordinary business of life (despite the occasional desire for the contrary). As Forsyth (2014) suggested of the use of camouflage for concealment, "[t]hroughout the Second World war camouflage was employed to conceal and counterfeit the military's presence in diverse landscape habitats; a technology designed to wipe out revealing patterns and inscribe false forms, all in order to make of enemy reconnaissance a glass eye" [125]. Hence human beings may detect other human beings and animals for companionship, for inter-personal communication and for the purposes of sharing information, in fact it is a feature of human relational intelligence that people have an ability to 'interrupt' the semiotic salience of an entity in relation to their environment, to as Leonie (2010) puts it,

". . . . represent a trace of the passage between observation of camouflage as a natural semiotic behaviour and the production of camouflage as a cultural semiotic technique" [126]. But another reason organism may use camouflage is that it buys them time of non-discovery in which to establish a relationship with the environment, it is a natural affordance as a consequence of belonging to a place, space and time.

Camouflage is thus adaptive for both humans and other organisms, it ensures that the semiotic relationship of the organism with its environment is on a continuum between contrast and invisibility, a camouflaged organism interrupts the relationships between the environment and the entity or object, from the perspective of the observant, and given the most successful combination the entity stops becoming a representation of a separate object and is perceived as simply a representation of the environment itself [127]. The patterns of camouflage also imitative of human foraging behaviour in patchy habitats (Wilke and Mata, 2012). When people (who have been hunters and gatherers for far longer than agriculturists) forage in high density resource areas surrounded by low density resource areas, they need to make decisions about both location and the time spent in patches as resources are depleted. Patch leaving behaviour is important, for as Wilke and Mata (2012) observe, ". . . the number of resource items within a patch can either be similar (evenly dispersed distributions), completely random (Poisson distributions)" and some may have few resources and other patches many [128]. Hence camouflage patterns may also correspond to the distributions, intensities and dispersals of foraging patch areas. As Marshall (2015) et al., state, "[s]election has typically driven better matching against local backgrounds" to enhance camouflage against predators among lizards for example but nevertheless there is not a clear association between the risk of predator and the degree of camouflage, rather the camouflage may simply by a property of the organisations belong to the environment [129].

For Webster (2015), there are two hypotheses for how disruptive camouflage functions, which are mutually exclusive [130]. The first of these is the 'local edge disruption' hypothesis which holds that camouflage is created or achieved by breaking up information edges in which an edge is defined as a curve or line that is the boundary point where the organism begins or ends and the background environment begins and ends. Thus it demarks the extent of continuity or discontinuity between surfaces. Edge elements are aggregated into shapes as a consequence of visual contour integration and completion [131]. Camouflage is attained by breaking up of edges making them discontinuous degrading the boundary between the organism and the environment [132]. The second hypothesis is the global feature 'disruption' hypothesis which states that camouflage acts so as to disintegrate the characteristic features of an animal masking its overall shape [133]. Having additional edge markings make an organism harder to recognise, when the eye-tracks an organism with discontinuous edge features broken-up outlines may be seen but dismissed. Consequently, a small increase in being seen and overlooked leads to larger gains in survivorship. Edge marking camouflage because the provide boundary concealment which is independent of the organism's shape. [134]. If there are a diversity of potential organisms to identify with inconsistent characteristics then the prey views them to reduce the distinctiveness of the edged; if there is only one kind of organism, the prey could benefit from concealing recognisable features [135]. Camouflage may be related to adaptations for aggregated environments, as animal appearances mirror some features of the environment form which they emerge. As Wilke and Rata (2012) observe:

> Species of plants and animals rarely, if ever, distribute themselves in a purely random manner in their natural environment, because individual organisms are not independent of one another: Whereas mutual attraction leads to aggregation for some species, mutual repulsion leads to regularity (dispersed environments) in others. Most often, these deviations from randomness are in the direction of aggregation, because aggregation offers considerable benefits such as a common habitat, mating and parenting, or the benefits of group foraging [136].

A camouflaged entity will achieve the effect of adaptive presentation of distributive markings, these come in three types. Firstly, disruptive marginal patterns which interfere with a boundary detection to mask outlines; secondly coincidental disruption in which high contrast markings on adjacent body limbs are aligned through behavioural posture creating a false boundary, that is more visible than the objects feature; and thirdly a disruption of surface which breaks up the uniformity of an object to make it appear disjointed. Disruptive objects are harder to discern and are frequently open to misclassification. An example is found in the disruptive pattern of the digital camouflage (see Figure 1). Humans and non-humans may share object recognition processes which are the perceptual conditions the entail the global feature disruption hypothesis and the fundamentals of object precognition involve edge detection, feature grouping and object classification. Bottom up processes provide texture and colour perceptions and top down the object related information [137].

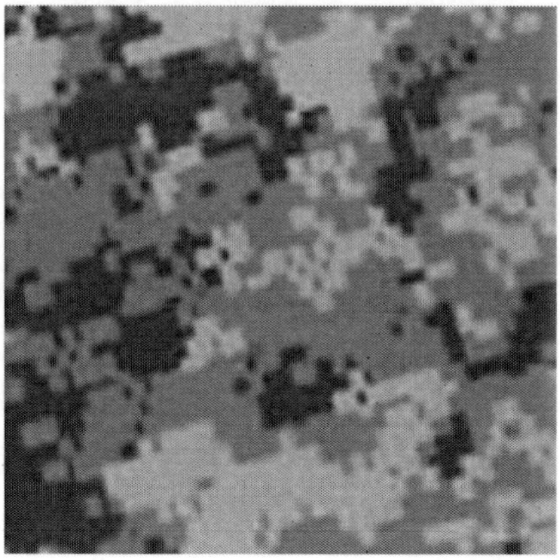

Figure 1. Pixelated or 'texture match' camouflage.

Different textures, shades, juxtapositions of colours and forms render the landscape a shifting canvas in which the animate is betrayed only by the contrast of motion that is not wind or wave, eyes ". . . adjust to the landscape's mauve atmosphere created by the interplay of sand, wind and sun, squint against the harsh, white light and peer more intently into the background to observe a group of soldiers" [138]. Conceived in the 1970s, pixelated or 'texture match'; camouflage composed of variously coloured and patterned squares remains the most effective form of universal military camouflage.

The efficacy of camouflage in preventing initial detection receives most interest despite the fact that predators often encounter the same phonotype, thus as Troscianko, Lown, Hughes and Stevens (2013) relate, "[t]he overall value of a camouflage strategy may, therefore, reflect an ability to prevent detection and to resist learning by the predator" [139]. Certainly, this might be the case for octopi and chameleons. Studies show that disruptive colouration are successful at concealment with high contrast patterns. But high contrast patterns had faster decreases of detection times compared to other stimuli and distractive markings may be learnt more slowly than background markings. On the other hand, high contrast markings were favoured when predators have limited opportunities to encounter same prey types and costlier when species are more abundant. Predators search for prey by a process of attentional priming but when detection becomes difficult they may use serial processes instead of parallel search. The success of the camouflage strategy is determined not just by how it evades predator sensory processing but also how it defeats perceptual and cognitive processes of predators in doing so [140].

The 'meaning' of the semiosphere must derive from the perception of objects and organisms in the environment but also in the way in which an organism veils, hides and prohibits its representation within an ecosystem [141]. It is an evolutionary advantage to detect a presence of others and hide the presence of self. As such organisms attempt to avoid recognition by either predators or prey [142]. But there is also a 'handicap' principle by means of which a sign suppression is abandoned in favour of a 'loud' semiotic presence, which signals that an organism is spotted but not 'off-guard' [143]. But this handicap principle might itself be an adaptation of a relational 'mating' strategy, connoting sexual aggression. Consider the display that a Peacock's tail makes, appearing both like 'large eyes' to potential predators and shimmering with attraction to the Peahen. In so doing predator and prey both avoid energy-consuming struggle [144].

According to Millikan (1984), internal representations of animals mostly consist of PPR (push-me-pull-you-representations) types which present one state of affairs and produce another [145]. Such internal mental representations do not just locate organism but contain the necessary information and affordances that describe what can entity do with it (attack, avoid, eat, hurt, mate with it) [146]. Hence there is a complex cognitive process of abduction, in which the organism infers certain facts that render inferences plausible, explain new phenomenon, and responds to schema in which hypotheses are formed and considered beyond 'instinctive' reaction [147]. A favoured inference will be one that leads to the survival of the organism. Not reacting to a predator is the best way to avoid being killed. Hence if a belief is false but successful, generating false beliefs may be adaptive by contributing to an organism's fitness [148]. Abduction can lead to misperceptions of affordances such as the false negative (for example calling an enemy a friend) or false positive (testing a disconnected wire for conductivity) [149]

Consequently, all organisms have inferential relevance that is both active and passive. This semiosphere of inference can disperse out into an environment and receive and process signs from organisms. It can also produce signs that fall out of prey and predators' agency-detection mechanisms but can be loud signallers. Elements of semiosphere reveal another organism to senses and its behaviour [150], "[t]he key lesson taught was that camouflage, in the desert, was not a benign technique for protection, it was becoming an ever more aggression and underhand method of visual warfare" [151]. A camouflaged organism is favoured by natural selection into modulating its footsteps and letting out signs which can be deceiving, or are meant to overwhelm an abductive threshold. Camouflage is aimed at preventing detection, avoiding recognition or preventing a PPR representation [152]. During states of crypsis, semi-abductive informational exchange is altered so that the organism attempts to minimalize the extent to which its argentic properties contrast with those of its environment [153].

Masquerade camouflage differs, it is not attempt at merger-camouflage but rather easily detectable but misidentifiable signs. Kinaesthetic camouflage is an alteration of a 'semiotic shadow,' which prevents an organism from being detected, and compromises quality of PPR representation. Some forms of camouflage are counterfactual; they are undetectable or overwhelming across an abductive threshold [154]. Camouflage can also operate in language, as no two person's minds can be completely transparent to one-another at any one time. People converse and set aside their differences in debates and arguments. Communicative camouflage involves the display of semantic performative acts which are likely to mislead interlocutors, shielding the intellect from actual beliefs of the truth or false variety. Some forms of sarcasm and hyperbole do the same. Conversants who are argumentatively opposed may state something they did not mean. This is termed *praeteritio.* Indeed, the strawman fallacy is an external sign aimed at puzzling and deceiving other interpretants [155].

Subliminal communication strategies are aimed at influencing a receiver's decision-making processes (to behave in certain way, or to buy certain products). It targets intense signals which stand-out from background which are below the threshold of consciousness [156]. The 'Lens model' is based on the idea that the relationship between the organism and the environment is mediated by so-called proximal stimuli, in which an organism can infer distal states of an environment [157]. In so doing the organism assesses the ecological validity of its semiotic traces. It is the quality of abductive inferences made by perceptual mechanisms within the semiosphere, which may counteract camouflage from improvement in cues in differentiating a potential threat. Biological camouflage acts at mimicking designs so that they do not alert agency recognition mechanisms, suppressing cues which allow differentiation [158]. Adapted for human use in conflict, camouflage offered a means from the earth through which to undermine and subvert the aerial view – to disguise the relationship between earth and air, either in the complex voluminous geographies of the battlefields, or in the habitats of competing predatory organisms [159]. Human dress in its various styles and shades, even away from the battlefield uniform, can be seen as a form of camouflage, a system of sign and status sharing within the myriad environments of country and city.

Chapter 5

SPEECH BUBBLES

Language is the basic medium through which people communicate. It is how people negotiate their world with one-another, converse, do work, form bonds and attachments, share thoughts and affection. As Pinker (1999) suggests, language is one of the main 'products' of the human mind and useful to our genes because it helps them to propagate, "[t]he ultimate goal that the mind was designed to attain is maximising the number of copies of the genes that created it" [160]. Between six thousand and eight thousand different languages are spoken in the world, far many more than are written down and for which there are distinct records, of which there are approximately only two hundred or so. From whence did language derive? The complex issue of language evolution continues to perplex people in communication, linguistics, archaeology, anthropology and cognitive science [161]. At the functional level, Mossman (2014) explains that language and other cortical functions such as 'consciousness' are derived from emergent properties which themselves are the result of "lower-strata neural networks" [162] But as we have and will argue consciousness and language are both learned, they occur on a 'platform' or 'plane' of understanding that is derived from the mind combining modular internal functioning with awareness of exterior information processing, language gives expression to thought which helps the body to mediate between it and the environment. So ultimately then it is learning and the capacity to learn which is the emergent property of the functional result of the complex interaction of lower-strata neural networks.

There are several opinions as to the evolutionary origin, or adaptation of the human species to an environment in which having a language function was immensely useful. These include that language originated in a gesture basis, in primitive vocalisations, or in song; but most probably a combination of all three contributed to human language development over approximately the last two-hundred thousand years since the emergence of anatomically modern humans (AMH). Language helps us to share ourselves, our thoughts, feelings and work patterns, it helps us to assimilate our environment and to organise ourselves with it. Evolutionary critics remind us that language, as much as technology is the pinnacle of human evolutionary achievement—culture and society would not be possible without language. As Harari (2012) states, ". . . our language is amazingly supple. We can connect a limited number of sounds and signs to produce an infinite number of sentences, each with a distinct meaning. We can thereby ingest, store and communicate a prodigious amount of information about the surrounding world" [163] While human babies are not born with immediate speech they are born with immediate expression and gesture, they are communicating beings, they learn to vocalise there sensory and cognitive thought patterns in communicative response to other human beings. Language is above all a social event. How is it that novice language learners can acquire a complex knowledge of language without necessarily knowing the grammatical rules and cognitive processes that lie behind it? Is language innate in humans? How it is that language gains a rate of change and complexity adaptation far above that of physiological adaptation?

As Fedurek and Slocome (2011) argue, "[l]inguists have been intrigued by this counter-intuitive gap, known as Plato's problem, between the quality and quantity of available linguistic data and the complexity of linguistic knowledge from which such inputs are created" [164]. Language seems to be so intrinsic as to be a bi-product of human life itself, having both generative and mimetic qualities. Indeed, people are 'made' out of a kind of language, a 'genetic code,' language might be seen then as an extension of the human body and mind. So how are humans programmed for language? As Jakobson (1970) suggested, the question is whether "the isomorphism of these two different genetic and verbal codes can be explained by a simple convergence stemming from a similarity of needs, or whether the foundations of the manifest linguistic structures, intimately based on molecular communication, are not directly patterned on the structural properties of the latter" [165]. We speak because evolution has made us to be speaking beings, we converse because it is a direct expression of human being, habitation and culture. Arguably every life form with a genetic code is part of a language and every living being that has agentic or expressive qualities speaks a language into the environment, whether or not it is conscious of doing so, or can consciously manipulate that language expression. Approaches to language acquisition and use have taken one of four paths: language structure and how structure is informed by human biology and physiology; the ability to acquire language among human and possibly some animal species; the functional conditions of language; and the history of the evolution of language [166]. Language is the communicative currency of the semiosphere, it is the oxygen (and carbon dioxide) of idea generation and sharing.

In contemporary times most theoreticians agree that language has arisen because of the interaction of three complex adaptive systems: biological evolution, human learning, and through the acquisition and development of human culture [167]. People use language to communicate with one another and to articulate their presence in the environment. However, biological evolution imposes physiological constraints, as does the human environment. Thus, conversely, it may be a poverty of stimulus which creates the need for a linguistic structure and a genetic and cultural substrate to linguistic development. We speak a language with a seemingly endless variety of expressions and propositions because we have an innate need to negotiate with each other our experience of the environment. Our language combines our sense into a focused expression of our being in the world, and at the same time it is a window on the mind. Like points of light which illuminate ideas or shadows through which we penetrate with our senses and thoughts, language expression is complex, and the rules and lexicons which govern its use are also constantly changing. The multivalency and ubiquity of language make it a distinctive property of human culture. As Fedurek and Slocome (2011) put it: "[l]anguage enables humans to represent and communicate complex abstract information, and it occurs in verbal, gestural, and written forms" [168]. Language is also inherently combinatorial, with syntax (sentence structure) leading to semantics (sentence meaning). Just as we are 'built' from the environment by a language of a genetic code we can 'build' with language an expression of our experience of ourselves in an environment.

The Functional Conditions of Language

In order to explain language acquisition, the linguist Noam Chomsky (1966) proposed that there is a sophisticated genetically encoded propensity for humans to acquire language, which comprises a Universal Grammar (UG) that undergirds language use, and a language acquisition device (LAD) that guides linguistic competence from observed and learned "social" data [169].

Some twenty years after the communication theorist Marshall McLuhan who proposed that the medium was the message, Robert Logan (1986/2004) asserted that there was an alphabet effect (which tipped the scales in the ancient world) [170]. Once oral language became written arguably civilisation had a form of vast analogue computer with which to order and structure the environment from small scale to large, it could record the effect of time's passing, and rearrange spatial coordinates that represented the physical properties of objects in the environment. However, the argument is made more complex from a variety of viewpoints, one being that writing might not be the "objective correlative" of speech but rather a way of changing language use, writing had a different property from verbal culture in that its recording required a further process than spoken articulation. Language changes constantly, and language from one cultures is sometimes absorbed into another, with population migration and the passing of time. Consequently, it could be said that the importance of the alphabetic tradition of ancient Greece became a product of European classical tradition, and thus of wider experience and education and culture [171].

Our cultural forms have been informed by the language and history and traditions of the cultures that have been absorbed through history into the present cultural experience. But not all languages are alike, languages can be comprised of different (though relatable) systems. For example, the differences between logographic (a logographic system uses a number signs representing a morpheme which is the basic unit of meaning, a logogram can be a word or part of a word), syllabic (a syllabic is a written character which represents a syllable or the nucleus of a syllable) and alphabetical systems (in which strings of characters are placed in order based on their position in the conventional order in a collatative process) are not entirely clear cut in as much as they all lead to phonetic signs (as do some forms of ideogrammatical writing). However, Grosswiler (2004) does point out that pictographic writing is highly stable within cultures and acts as a force of cohesion (a kind of cultural bond), for example, "because Chinese has not been affected by phonetic changes, dialects, or linguistic structures, it has been a means of political unity since the end of the third century BC" [172]. So a culture which does not share a common stable language may not share as strong a social bond. This raises the question of whether language always meets the needs of its speakers: can an idea exist if there is no word for it? On the face of it this argument does not contain much sense, in so far as it cannot account for the mental life of people who have greater or lesser vocabularies, or for the use of the imagination and other creative insights to produce new combinations of existing ideas.

But as people evolve so too does language. However, cultural conditions shape language growth and decline. As Starte (2006) suggests, "there is no absolute distinction between the evolution of media, language, or that of the human species" [173]. We cannot conceive of human society without forms of communication and language. It is also often said that "language does not fossilise," but the first forms of recorded expression are thought to be those that co-evolved with technological innovation and took place between some twenty thousand and thirty-thousand years ago. Starte (2006) also states that there was a progression from logographic and phonetic writing, to forms of syllabic and alphabetic writing [174]. What makes language special is that, although the vocabularies of most language users' vocabularies are limited to fifty-thousand words or so, those vocabularies are nevertheless constructed in a way that allows for unbounded transmission of messages which, when undergirded by a universal grammar, can result in a high fidelity of communication between language users. However, cultural patterns such as those brought about by tourism, migration, commerce, and corporatisation, can lead to forms of homogenization under common global conditions which can erode the variety of languages spoken. It is generally thought that language variety is in decline under conditions of modernity unless steps are taken to reverse this process, through such means as language nesting and revitalisation. Thus the recording of language systems, the vast variety of which are largely oral, is crucial to maintaining the currency of cultural differences. Language shapes thought, and different language systems may shape some forms of thought differently. The grand and vast augmentation of human being and conscious thought which is encapsulated in language systems and their use is too precious to be blunted by linguistic erosion when this process could be forestalled. So language is very much an extension of human thought at the cultural level, but at the physiological level it seems it is an extension and expression of who we are also.

From a bio-linguistic perspective, Berwick and Chomsky (2011) state that:

> …we can think of language as, in essence, an "organ of the body," more or less on a par with visual or digestive or immune systems … it is a cognitive organ, like the systems of planning, interpretation, reflection… We can think of language as a mental organ, where the term "mental" simply refers to certain aspects of the world, to be studied in the same way as chemical, optical, electrical and other aspects… [175]

The use of language is predicated in humans on the expression of the FOXP2 and CNTNAP2 genes and the population frequency of two brain growth and development genes, ASPM and Microcephalin [176], but the array of possibilities for language use is limited only by cognitive complexity. However, fortunately the human brain at approximately 1350cc has evolved just for the management of such complexity, and in the human brain the temporal lobe is twenty-three per cent larger than other brain regions, and four times larger than that of our closest mammalian relatives, chimpanzees [177]. Linguists and cognitive neuroscientists have also determined which regions of the human brain are involved in which parts of speech with some degree of precision. For example, the understanding of proper nouns is thought to occur in the anterior and medial areas of the temporal lobe, and that of common nouns on the lateral and inferior temporal lobes [178].

Thus language is enabled and constrained by human biology and physiology. The levels of human physiology that enable language stem from the brain and reach to the organs involved in the production of human speech, as well as those for the processing of the speech of others. At a higher level in language semantics than the functional structural underpinning of universal grammar is the language lexicon, which may vary in arbitrary ways from language to language, allowing language history to be traced by shared lexical commonalities. Language can thus be thought of as a complex adaptive system that comprises a series of interacting agents, each of which responds to local conditions following simple rules [179].

While many species on the planet communicate, the cognitive possibilities that spoken and written language affords are thought to be distinctively human. It is unlikely that non-human animals or indeed any other species has syntactical grammar rules as complex as humans. As Fedurek and Slocome (2011) put it: "[h]umans can convey an infinite amount of messages using a limited number of words because of the powerful system of grammatical rules that govern the structure and form of language, including the ordering of words into meaningful sentences" [180]. As such, language allows people to divide the external world in a myriad of ways and to create artificial worlds, such that it is nearly impossible to imagine a social system comprising human beings that is not ordered by language [181].

The anthropologist Leach (1982) argues that the transmission of culture, in the sense of patterns of learned behaviour, from generation to generation by learning instead of by genetics, is a characteristic of many other species: "[b]ees and ants have extremely elaborate, highly organized social systems which appear to get along without any concept-forming medium of communication which is even remotely similar to that of human language" [183]. Thus it is clear that non-human languages exist, and that language itself is necessary property of human culture or indeed of animal culture among some species. So just as with human beings it makes less sense to think of a bee as being a discrete animate object than tied in relationship to other bees, through its complex iteration of flight dances to reveal food or 'home' locations, just as we are tied to other humans through our language use, to negotiate our food and shelter.

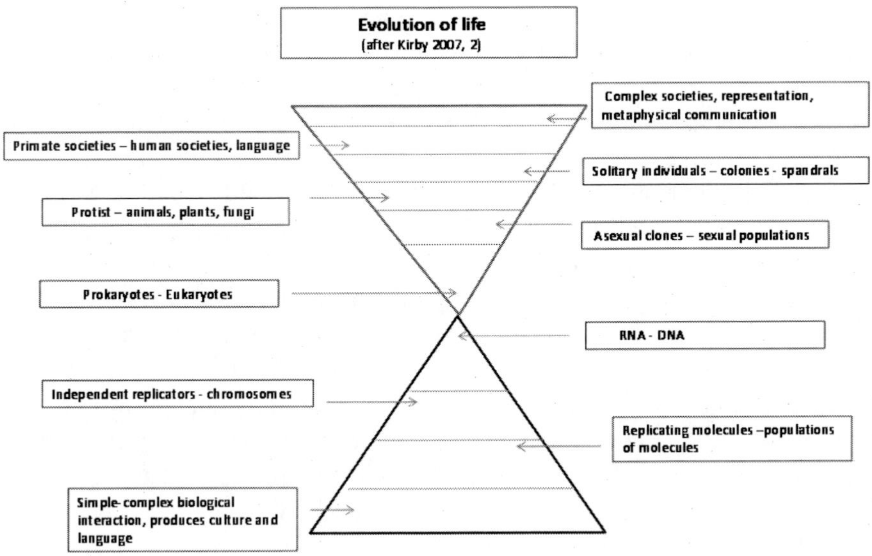

Figure 2. Evolution of life (after Kirby 2007, 2) [182].

How Language Is Informed by Biology and Physiology

Hauser et al., (2002) argue that language as a biological faculty may be divided in a "Faculty of Language in a Broad Sense" (FLB) and a "Faculty of Language in a Narrow Sense" (FLN): the former may be analogous to animal communication and include aspects of language faculty (includes the perceptual-articulatory apparatus); and the latter may be analogous to the human language faculty (and includes the conceptual and intentional apparatus) [184]. Kirby (2007) [185] and Corballis (2009) [186] also assert that recursion is a defining property of human language. Recursion is defined as a procedure in which one of the steps involves invoking the procedure itself, such as that involved in nesting syntax meanings within sentences. However, others hold a view that a reliance on recursion as the main feature of human linguistic difference may oversimplify syntactical aspects of human language, in which communicating propositional structures is an adaptation consisting of many different interacting sub-systems within human consciousness and physiology. Thus, if there are common features between FLB sense and FLN, then the process of complex language use involving recursion, intentionality, empathy and meta-representation is a gradual one involving intermediate stages that also have evolutionary correlates and limitations. After all, the most distinctive property of language is that it is generative, allowing parts of language syntax to be recombined to make other language structures, which involves the "embedding and concatenation of phrases" [187]. Consequently, is language use, like evolutionary behaviour, a process of constant adaptation? Is there a plausible evolutionary trajectory from protolanguage to ways of communicating that have yet to be imagined? Possibly the generative properties of language could be due to empathy producing processes that are the consequence of learning enabled by so-called 'mirror-neurons.' As Lacoste-Badie and Droules (2014) state: "[m]irror neurons are a particular class of visuomotor neurons in the brain that show activity when an individual performs an action and when (s)he observes another individual performing the same action" [188]. Hence the linguistic mapping associations produced by the exchange of language sequence are conducive to empathic exchange. True or not, the human capacity for empathy is one of the great unexplored communicative possibilities.

As Richardson and Boyd (2010) observe, the division of labour between genes and culture is a co-evolutionary process [189]. The Baldwin effect argues that language adaptation occurring within an individual's lifetime may affect linguistic competence: in other words, if language use and linguistic competency gives an advantage, a trait change within an organism's lifetime as a part of cultural evolution can be assimilated into the epigenetic repertoire. As Richardson and Boyd (2010) explain, "so long as a higher capacity cultural communication system was favoured, cognitive modifications to more efficiently acquire explicitly linguistic features like symbolic words and grammar would be favoured as the system passed some threshold of complexity" [190]. Thus it is possible that a Baldwinian effect[10] can lead to

[10] The Baldwin effect concerns learned behaviours. An organism's ability to learn behaviours effects its reproductive success and hence the genetic makeup of the species. The difficulty

random selection for linguistic competencies if they provide an adaptive advantage. In one sense, then, biology and social evolution act as a 'fixed-point attractor' for human linguistic competency [191]. A 'fixed–point attractor' is a patterned system in complexity theory and chaos physics in which an attractor is represented by a particular point in a phase state termed an equilibrium point which corresponds to a limited range of possible behaviours, for example, a pendulum at rest. However, as Yamauchi and Hashimoto (2010) point out, it is possible that a strong assimilatory process might be available within ten to twenty generations with a selective environment that is dynamically niche-constructed, but a functional redundancy may result on the genetic information of the learned trait [192]. It is possible that the redundant genetic information becomes degraded while no effect is observed on a linguistic level. For example, the gene for birdsong results in natural selection of related adaptive physiological changes such that a functional redundancy occurs without externally motivated environment changes [193]. Arguably, the exciting things about language use and birdsong might happen in just a cycle of gene recession as other genetic factors are selected. For example, syntactic recursion may facilitate more complex language use that results in a gradual redundancy of selection but an increase in a selection of genes for other language features such as intentionality, with corresponding physiological adaptation in the neural networks.

If language is a form of cultural evolution, it is useful to look at the relationship between its historical progression and changes in human evolutionary physiology. Whilst some observations have been made (such as the descended larynx that enables human speech), it is difficult to isolate any one physiological feature in cognate language use (red deer and lions, for example, also have a descended larynx). More probably, language production is enabled because of the interaction of a number of different musculo-skeletal and neural features of human physiology, some of which are shared by other species. Thus language could evolve from a rudimentary set of culturally transmitted signals, impoverished vocalisations or manual gestures to symbolic words and grammar via selective genetic adaptation, given the relative social utility of such communication. However, people are the product of their genes, gene expression, and also the environment—all three are involved in the evolutionary process. Cultural evolution leads to the adaptation

with the Baldwin effect is that it applies to group contexts. Concepts such as altruism (which are seen as morally positive and culturally learned) may lead to different consequences at an individual level. The sustained behaviour of a group can be a shaping influence on the evolutionary process in a non-determined way.

of a universal grammar. We can ask then the fundamental questions: What is the purpose of human language? Why do we not all speak a common language? What is the reason for linguistic diversity?

Richardson and Boyd (2010) suggest that linguistic diversity arises not just because of "cultural containment" by ethnicity or geography, but also for the sociological reason that it limits communication between people who cannot freely trust one another or in which truthful communications lead to maladaptive behaviours [194]. However, this is increasingly tested by conditions of globalisation, in which it could be argued that technology precedes the use of language adoption. If global societies and economies are adaptively integrated, then the adoption of one language might be useful. In this way humans are also different from some other animal species in depending on social transmission for adaptive behaviours.

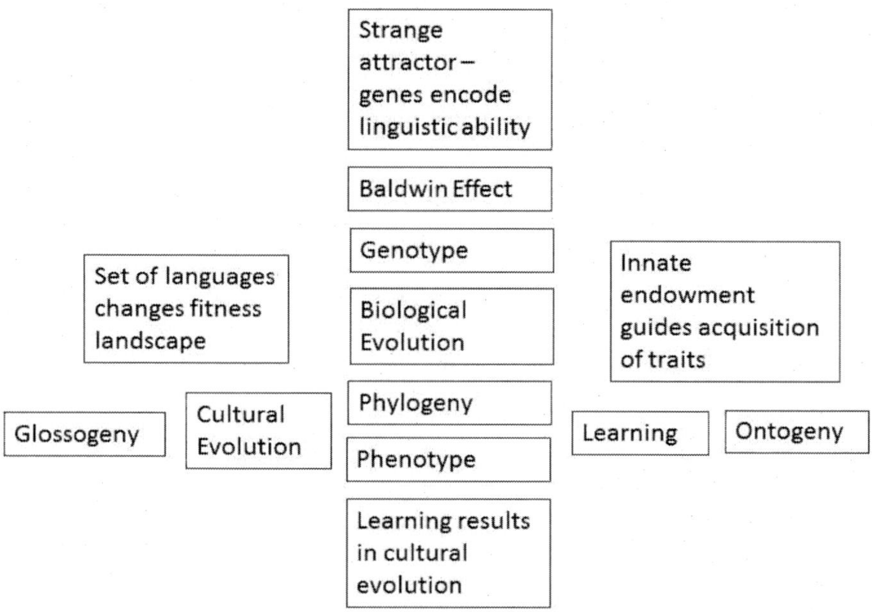

Figure 3. Three-cycle process of language change [195].

The Uniqueness of Human Language

Intentional agency enters into most human communication. Because of the innate complexity of language, individuals may not have enough in common to facilitate open and honest communication all the time. Furthermore, individuals and societies are competitive and consequently many people have mixed motives in communication, including self-interest. If not all communication, speech and language is trustworthy, then some features of language may not have a selected advantage for evolution, depending on the cognitive and social cost of information exchange. This leads to two features of human language use. Firstly, language variation may be adaptive; and secondly, because combinatory communications systems have rules of interpretation, they may be vulnerable to both misunderstanding "noise" and also to deception. Thus it requires concerted forms of co-operation to enable language evolution. Nevertheless, language diversity may not have a basis in issues of trust. Presumably niche-adaptation is the most likely cause of such diversity, not deviant inventiveness. However, human biological complexity and conscious cognition in language occur on related but different levels of affect. Human language undoubtedly arose from adaptation through natural selection from principles of imitation and sexual selection [196] and the self-reinforcing quality of language use and inter-generational language learning infers that "information transition by memes is faster than that by matter as in the genetic process" [197]. Thus it is the ability to communicate through language that empowers the human societies as a social organism far beyond that of other species. However, there is no precise border on the continuum between humans and other species.

Coeval with language use, people have the ability to represent different referential locations of time and space through symbolic communication. While the basis of that ability might lie in gene acquisition and physiological capability, the sociality of language is largely separate from biological substrates. Hence we can imagine that language use evolves in an individual's lifetime through acquisition of different linguistic competencies: for example, grammar and lexicon (tens of years). This stage is known as ontogeny, possibly leading to genetic adaptation through the Baldwin effect. However, languages evolve over hundreds of years (think, for example, of the differences in vocabulary between Chaucer, Shakespeare and the modern novel). This stage is known as glossogeny. The ability of species to engage with language abilities (such as a universal grammar) are acquired over thousands of years. That is known as phylogeny. MacWhinney (2005) adds an epigenetic stage (a combination of environmental and genetic interactions), a developmental (ontogenetic) stage, a processing time-frame, a social emergence time-frame, and a diachronic (glossygenetic) period of language change [198].

However, as Dediu (2011) suggests, "it is becoming increasingly clear that not only are the pathways connecting genes to phenotypes nonlinear and difficult to map, and that gene-gene and gene-environment interactions are the norm in the development of most phenotypic aspects, but also that there is no clear-cut difference between 'genetic' and 'environmental,' that 'development' is not a discrete, encapsulated, and teleological phase in the life cycle of an organism, and that 'genes' are essential to all processes at all times" [199].

It is difficult to tell from evolutionary archaeology exactly when changes in human phylogeny resulted in the genetic selection for complex language ability amongst humans. Fitch (2010) suggests that it occurred somewhere between two-hundred thousand years ago and the present day [200]. In most people the control of language is co-ordinated by the left brain hemisphere—the regions of the brain which activate and integrate motor control of the lips, tongue and vocal cords. It is the claim of many linguists that human beings are born with a knowledge of linguistic structure and the ability to acquire language (known as a Universal Grammar), which is encoded in the genome at least in part by the FOXP2 gene, which is linked to the linguistic features of the brain found in Broca's and Wernicke's areas and the pre-frontal cortex. The proponents of Universal Grammar argue that people are born with linguistic competency which has a genetic substrate. Babies are not born talking but they are born with an 'innate' ability to form linguistic structures. However, relatively small changes at the phenotypic level can lead to quite large cognitive and behavioural changes.

Comparison with other animal species reveals that some animals (such as dolphins and parrots) have quite sophisticated imitative skills, which may imply a form of language-related empathy; similarly, animals may have "functionally referential" alarm calls capable of bearing external referents. Apes are also capable of audience effects that can bear "conspecific" information. However, as Fitch (2005) suggests, many such calls are not intentionally referential and might not shape calling in ways relevant to the listener's knowledge [201]. This does not necessarily imply that animals cannot individuate at a mental level in their communication, but rather they may not have theories of mind that are capable of attributing more than physiological awareness of others. Thus many animals lack an ability to "shape" their communication in ways that humans do. Hence, language (if not symbol manipulation) may be an exclusive property of the human species which has evolved gradually over several hundred thousand years. Ekstig (2015) argues that human language development is a form of disruptive natural selection, with nothing preventing the evolution of higher complexity (due for example to competition from more complex language users) [202], as he states, "[o]ne may say that language made the brain visible for natural selection" [203]. Hence language may seem to be another 'sense' or an extension from a combination of senses and cognate structures. It plays a part in structuring the expression of our thoughts and feelings (whether objective or subjective), and it is means by which the world becomes known to us.

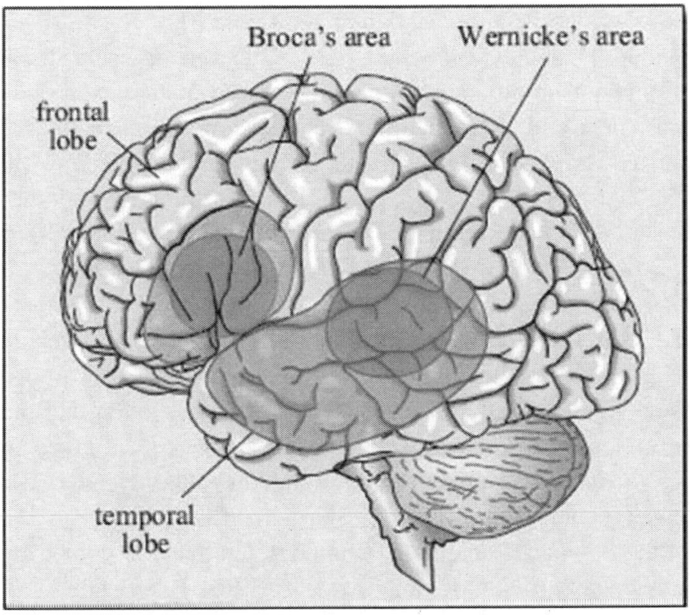

Figure 4. Diagram of the human brain showing language function areas.

According to the argument of language evolution as a gradual adaption in human society from genetic assimilation and acculturalisation, the underlying genetic substrate of language enables a Universal Grammar, which is a set of grammatical principles that applies across all human languages. This goes some way to explaining language diversity—each society develops its own adaptation to a different linguistic environment that fits its locality and social identity. However, language might be shaped by physiological limitations that are not entirely dedicated to language. These include perceptuo-motor factors, including physiological features that account for the seriality of vocal structures, and perceptual systems that store sensory information (involving the retina, primary visual cortex, occipital lobe, the dorsal stream and the ventral stream; the vocal tract affected by the tongue, whose muscles are controlled by the hypoglossal nucleus and nucleus ambiguous); and lower jaw (controlled by the trigeminal nucleus; and lips controlled by the facial motor nucleus). Secondly, language might be shaped by cognitive limitations on learning and processing, including memory, and from structures of mental representation and reasoning (involving the cerebellum, lateral hemispheres, Broca's and Wernicke's areas (see Figure 4), the *arcuate fasciculus*, and pre-frontal cortex) [204]. In order to describe the two inter-related issues of language development, Chater and Christiansen (2009) have specified the terms "N-induction" and "C-induction." "N-induction" describes the language challenge of understanding and manipulating representations and relationships in the natural world, and "C-induction" describes how people acquire ability to converse and co-ordinate with each other. They make the point that in N-induction the world imposes an external standard, but in C-induction the standard is social. The difference is that in the latter language conditions people to do the same thing but not necessarily the objectively true thing. While both conditions are predicated on learning, the second does not rely only on the correspondence theory of truth but also on "relational" or performativity aspects of language. Chater and Christiansen (2009) make the point that language is mostly acquired by C-induction, given that there is no human-independent true language which people learn [205]. Rather, they learn the language of their developmental environment (although second-language learners may have more than one such environment).

Thus in the languages adaptionists' model, language is guided by genetic inheritance and the growth of cognitive structures (phylogeny), which are shaped by the environment as the learner acquires the language of their caregivers and culture, allowing for language evolution at the level of glossogeny. Such language acquisition contributes to the reproductive potential of the individual. At the level of individual and group language use, language may be seen as a mapping device between meanings and signals. In this way languages may be composed with a shared signal structure, or be seen as being holistic, whereby such structure may or may not be implicated in a shared meaning space [206].

Private Language

There are two main theories of language acquisition. Basically, the first theory posits that all language must be learned; the second theory (currently most favoured by linguists) is that language is an abstract system that fundamentally cannot be learned without an innate language faculty, or by learning which is achieved on top of the genetic inheritance of a substrate of universal grammar. Some linguists posit that humans have evolved with an innate propensity for language and that certain syntactical faculties such as the ability for symbolic recursion are 'hard wired' into the brain and human physiology. The study of meaning in language (semantics) also has a connection to psycholinguistics given that letters and words, and sentence structures, are forms of symbolic description: "symbols are thus used in models of cognitive processes (internal symbolic representations) and in models of natural language (external symbols linked to internal symbolic representations)" [207]. The symbol for a banana, for example, may be similar for both monkey and human.

This tends to lend weight to the refutation of the private language argument in Wittgenstein philosophy that although Words not causally related to sensation but arbitrarily assigned – sensations can trigger words for the experience of them, nevertheless, all of the elements of a language must be publically available. Thus the meaning of words is not fixed by reference to our private experiences, and so data from subjective state claims indicates nothing about or subjective experiences, for if a grammar is universal there must be a universal code for it grounded in genetic substance. As Kuczynski (2005) suggests, knowledge of a language involves a grasp of propositions, where there is no grasp of propositions and hence no intelligent thinking, there is no language [208] Grasping a proposition as a rule following exponent of language is to have 'certain dispositions' [209]. There is a difference between obeying a rule and breaking it – a rule permits only its function and furthermore, a private rule is an unenforced rule for which there is presumably no difference between obeying and breaking – hence the idea of a private rule is unintelligible apart from the rule being a part of public practice [210]. Semantic rules cannot be preserved in private language hence several people are required if a language is to be known [211].

Whilst a universal grammar might be promulgated by social interaction and learned experiences it is nevertheless distinct from a private language rather constituting the genetic propensity to be able to from a language that can be socially understood. As Floyd (2006) states, it is "difficult to talk of a private sign 'S' having any linguistic role whatsoever that is not grounded in public roles" [212]. However, the criterion for correctness is reliant on memory and a mnemonic device 'as if someone were to buy several copies of the morning newspaper to ensure him or herself that what was said was true' rather than relying on the semantic value of the statement applying to a variety of valid contexts with regard to the content of the statement. As Floyd (2006) suggests, ". . . sensation words have to tell us something about what kind of sensation they're referring to, otherwise it would be difficult to see any difference between 'pain' and 'pleasure' " [213]. However, as Meyers and Waller (2009) state, the strong interpretation of the private language argument holds that language must be publically available, thus no meanings are 'fixed' in reference to private experiences and data that is available from perceptions of subjective states may not necessarily tell us facts about subjective experiences [214]. This is not necessarily to refute the claim that linguistic usage involving subjective states don't have correlations between brain states and subjective states but rather they are not causally connected through private language [215].

Social rule following in linguistic behaviour is rather a better candidate for language acquisition; such rules may be predetermined by a propensity for universal grammar in the Chomskyan sense. This is supported by Norman (2000) who suggests that reasoning is a social practice, a reason fit one person may be fit for anyone, hence reasons are public imperatives which may or may not be interpersonal, linguistic meaning requires an independent standard of reference so too does legislating for oneself. As Norman (2000) explains, "[j]ust as we cannot help responding to other people's linguistic utterances as meaningful; so we have moral obligations, which require us to value the humanity of others. The public character of reasons is grounded in our social nature." [216]. However, just as Norman (2000) affirms that any agents' need for autonomy or security is a reason for them to act, it does not follow that an individually would universally affirm that in others, the public reasons argument for language use and rule following "does not give other-regarding reasons any privileged status." [217].

Thus language is enabled and constrained by human biology and physiology. The levels of human physiology that enable language stem from the brain and reach to the organs involved in the production of human speech as well as those for the processing of the speech of others. At a higher level in language semantics than the functional structural underpinning of universal grammar is the language lexicon which may vary in arbitrary ways from language to language allowing language history to be traced by shared lexical commonalities. Language can thus be thought of as a complex adaptive system that comprises of a series of interacting agents, each of which responds to local conditions following simple rules [218].

While many species on the planet communicate the cognitive possibilities that spoken and written language affords are thought to be distinctively human. It is unlikely that non-human animals or indeed any other species has syntactical grammar rules as complex as humans. As Fedurek and Slocome (2011) put it, "[h]umans can convey an infinite amount of messages using a limited number of words because of the powerful system of grammatical rules that govern the structure and form of language, including the ordering of words into meaningful sentences" [219]. As such language allows people to divide the external world in a myriad of ways and to create artificial worlds, such that it is nearly impossible to imagine a social system comprising human beings that is not ordered by language [220].

Chapter 6

DÉJÀ VU

The uncanny feeling of heaving experienced something before. The brief sensation of being inserted between parallel universes. The trace of memory in the absence of recall. A momentary 're-setting' of cognitive disposition, a fleeting side-effect of information processing. This chapter examines the theories of the causes of déjà vu or stimuli "recognised as familiar without recognition of the basis of their familiarity" and posits that déjà vu may be an adaptive function concerned with information processing in the cerebral cortex [221]. We explore a psychological 'profile' for the experience of déjà vu and draw inferences about the physiological 'purpose' of déjà vu. At its most fanciful déjà vu experiences may lead to people believing themselves to be living in a computer simulation [222]. Explanations of déjà vu or the perceived feeling of "any subjectively inappropriate impression of familiarity of a present experience with an undefined past" [223] is a common lifetime experience among approximately sixty-seven of the population [224] and "occurs about once a year in experients" [225]. It is thus consistent with psychological normality and may in fact be a cognitive adaptation for processing new stimuli.

It is possible that déjà vu is a sensory experience that registers an adaptive evolutionary function of reorientation and 'resetting' of cognitive experience, similar but far briefer to the kinds of processing that occur during sleep. Whilst it is not possible for evolutionary psychologists to delve back in time and conduct investigative interviews with humans at different stages in evolutionary development over the last million years to see if they experienced déjà vu (how would an experimenter conduct oneself with a Neanderthal group, for example?) but imagining a time machine made such inquiry possible a present day psychologist would register the fact that the different branches of the humanoid family, most likely experienced the sensory world and the cognitive world somewhat differently from modern humans. There is some evidence that Neanderthal experience of colour, for example, was more vivid than present-day humans, and other hominid forms may have had more acute senses of smell. Language ability is enhanced in modern humans and phenomenon such as déjà vu is associated with memory recall and hence may be only experienced within homo sapien physiology with its larger memory store in the neocortex.

It is generally believed that human memory falls into three categories. Firstly, there is procedural memory which accompanies skill acquisition and may not involve conscious recollection [226]. Secondly, there is semantic memory which involves the recollection of facts (not unique to humans). Thirdly, episodic memory involves the recollection of specific events [227]. Only humans are thought to be capable of episodic memory and it is this memory dimension which most concerns the experience of déjà vu.

From the psychological perspective O'Connor and Moulin (2008) identify two types of theory classifications for déjà vu. Firstly, progressing from small or subordinate units involving perceptual theories arising from familiarity invoked from a perceptual environment, and secondly, theories suggesting it results from an 'over-arching cognitive feeling' that is applied to a perceptual input [228]. Even experts describe the phenomenon as 'a glitch in the working memory of the system' or 'a problem with how the perceptual world is communicated to the part of consciousness which is interpreting it' [229]. However, from the perspective of cognitive processing, déjà vu occurrences within the normal range of experience may be useful. Given that most peoples' experience of déjà vu is double-edged – an experience of temporal illusion that is 'recognised,' this may serve to re-orientate the experient to a conscious state that is results in an enhanced perceptual awareness. This could be a functional adaptation in modern humans that is perhaps on the same phenomenological dimension as the apparent memetic empathy of 'yawning' when tired, to enhance oxygen uptake to the brain.

Similarly, déjà vu has been described as paramnesia of 'wrong recognition' of an experience [230]. However, this is paradoxical as it is really the 'right recognition' of a 'wrong experience' and hence very useful for realigning cognitive processing with both subjective and objective reality. As O'Connor and Moulin (2010) suggest, "déjà vu is . . . a benign experience, not a pathological one, and does not lead to a behavioural impairment" [231]. The experience of déjà vu, 'an experience of familiarity without identification of their source' occurs on a continuum of 'recollection based recognition' [232].

Strong familiarity signals are produced by a high degree of salience between features of a current situation and recollections of previous experiences in memory whereas a low degree of overlap produces a weaker cognitive signal. Thus similarity of experience to previously experienced situations produce a feeling of familiarity, familiarity increases with resemblance (feature overlap) between situation schema experienced in the subject present and those that are stored in memory [233]. Déjà vu occurs when familiarity is experienced (feature overlap) without the possibility of temporal coincidence of perceived reality. Furthermore, in the déjà vu illusion the source of the feeling of familiarity may not be identifiable. This offers us the first clue to identifying the experience of déjà vu – bearing in mind that déjà vu occurs when we experience a feeling of familiarity despite evidence to the contrary, this experience may be limited to those in which there is an overlap between, a.) a feeling of familiarity, b.) an inability to identify the source of familiarity, c.) evidence that objectively the event could not have occurred before. Thus, although almost instantaneous there may be a temporal dislocation that precipitates the déjà vu experience – a feeling of prior experience that is a minute displacement of time perception in cognition of present experience. From this perspective, déjà vu is thus a learning experience – an aid to objectivity. Most of the literature points to an irregularity of phenomenon which may or may not have similar causes.

As Brown states, about sixty-per cent of the human population has experienced déjà vu. However, its frequency decreases with age, it appears to be associated with stress and fatigue, and shows a positive correlation with both education and socio-economic level [234]. Most surveys of déjà vu only ask about incidence rates rather than the qualitative dimensions of the experience. A survey conducted by Kohr in 1980 found that fourteen-per cent of respondents had one or two lifetime experiences, ninety-per cent had three of four, twenty-three-per cent had five to eight and forty-four per cent had nine or more [235]. The inference from this might be that if you have one experience of déjà vu you are likely to have another, indeed, as Brown claims, ninety-eight per cent of those who have experienced déjà vu once are likely to experience it again [236]. However, due to the relatively intense qualitative experience of déjà vu, it may be that the 'familiarity effect' is reinforcing, respondents are more likely to report more incidences than those actually experienced.

Although déjà vu may be associated with mild stress and anxiety, according to Neppe (1983) it is not accompanied by changes in thinking or emotion. Both Neppe (1983) and Brown et al., (2003) found that the déjà vu experience is triggered by a visual scene and is very brief in duration, the internal reaction is one of surprise, and involves a sense of temporal dislocation – the experience of time as slowing down [237]. However, it is very difficult to 'repeat' the 'scenario' for these experiences in experimental conditions. Thus many features of the déjà vu experience from the experimental position remain relatively unknown, there is either little data available of qualitative experience or that which exists is taken from very small samples (Neppe 1983) [238]. From the perspective of experimental design this is partially due to limitations of 'drawing information from retrospective designs and the fact that from a qualitative perspective one's memory of the experience of déjà vu is more likely to be dominated by the perceptual experience of it rather than objective knowledge of one's physical, psychological or neurological state.'

What is consistent from the studies which have been conducted is that the incidence with which it is experienced decreases with age [239]. Studies have shown that it is more prevalent amongst younger people [240]. Why is this? However, there are several possible further reasons why déjà vu is more common in twenty-twenty-four year olds. The first of these is neuroplasticity – certain sections of the brain are still developing at that age, leading to more adjustments in speed of neural transmission. It may be that the developing brain is less familiar with processing new information and so overcompensates in seeking experiences of familiarity – resulting in déjà vu experience. This could also be in part be influenced by socio-cultural factors. From the socio-cultural perspective, younger people seek novelty of experience and this expectation results in inconsistencies in neurotransmission as the brain adjusts to process both new and familiar situations.

A study conducted by Chapman and Mensch (1951) revealed that better educated persons had a higher déjà vu incidence than less educated persons. These results were replicated by Richardson and Winokur in 1967 who found a higher incidence of déjà vu in professional and student groups (forty-seven per cent to seventy-three per cent) than unemployed and unskilled groups (twenty-five perecent to forty-three per cent) [241]. Perhaps this is because better educated people may be more specialised in their common information perception experiences, variations of information difference have a smaller range thus changes in cognitive processing may be more pronounced in conscious perception. Or it may be that this generally narrower range of perceptual experience of information occasionally and momentarily produces over-excitation of specific regions of synaptic activity. There is also evidence that those who travel more may have a higher incidence of déjà vu because of the fact that they encounter more physically new locations than those who don't [242]. The study from 1951 by Chapman and Mensch (1951) revealed that those who do not travel have a eleven-per cent incidence of déjà vu, people who make from one to four trips have a thirty-one per cent incidence and those who travel more frequently have a thirty-two-per cent incidence of déjà vu [243]. Clearly, travel may be a stimulant of déjà vu but given that déjà vu is a phenomenon of historical commentary in times of far less travel, implies that this doesn't by itself, represent a causal determinant. Titchener (1924) also found that déjà vu may also occur with physical or psychological distress or in situations of mental fatigue [244]. The déjà vu experience may also be related to memory of dreams. People who remember their dreams are more likely to experience déjà vu.

Dream memory is also thought to be a sign of mental health, so clearly déjà vu correlates with normal experience. It may even serve to enhance normal experience by reorientating the individual to their subjective or objective present experience – it is a servant not a slave of consciousness. Thus the three components of travel, dreaming and repeated visual stimulus may result in potential sources of 'familiarity' stored in the memory of people who experience déjà vu. If familiarity is accompanied by an inability to retrieve the source, it leads to a sense of having been already experienced [245].

EXPLANATIONS FOR DÉJÀ VU

According to Brown and others [246], Neppe (1983, 2010) [247], explanations of déjà vu fall into four categories. These are: Fisrtly, dual processing (two cognitive processes that are momentarily out of synchrony); secondly, neurological (disruption in neural transmission or seizure); thirdly, memory-based (implied familiarity with unrecognised stimuli); and fourthly, attentional issues (unattended perception followed by attended perception) [248]. Among scientific studies there is little data available about the precise nature of the déjà vu experience [249], this may partly due to the fact that the experience is qualitative, hard to locate in neurological terms, fleeting, not inconsistent with normality, difficult to describe or replicate in quantitative experiments.

Whilst there is agreement amongst researchers that déjà vu is a 'routine cognitive experience unrelated to severe psychological disturbances,' we can openly speculate that déjà vu is not symptomatic of 'error' in cognitive processing – it may in fact be a very useful response for 'objectivity normalisation' – a functional adaptation.

Dual-Processing Explanations

The dual-processing explanation asserts that there is a disruption in the operation of two separate but interactive cognitive processes. Mnemonic processes which operate in concert may become momentarily asynchronous with one being activated in the absence of the other [250]. Thus familiarity and retrieval are independent cognitive activities usually operating in synchronous parallel hence recall is accompanied by familiarity of information retrieved. However, in déjà vu, retrieval can be activated in the absence of familiarity to become momentarily unfamiliar (jamais vu), or familiarity may be activated in the absence of retrieval (déjà vu). This dual process interpretation is supported by the earlier work of Bergson – perception and memory are simultaneous events "memory is never posterior to the formation of perception: it is contemporaneous with it. Step by step, as perception is created, the memory of it is projected beside it, "as the shadow falls beside the body" [251]. However, this theory may fail to account for the 'persistence' of memory, why one any given memory selection may be retained over another – consequently the differences between short and long-term memory and the selective filtering and 'assembling' of perceptual experiences within cognition.

It remains an enduring fact that people usually explain cognitive phenomenon in terms of the technology of the day, so extending this metaphor from the work of De Nayer (1979) we might speculate that under normal conditions, memory encoding and retrieval operate similarly to an mp3 player, either on 'play' or 'record' but not both at once [252]. Consequently, they can both be briefly active simultaneously – new experiences can briefly be encoded as familiar. However, Pashler (1994) has suggested that there is instead a 'bottleneck' model of attention, with a single process for memory encoding, memory retrieval and response selection [253]. Consequently memory and encoding may not work in parallel, and memory is the outcome of whatever processing occurred as input.

Other theories of dual-processing glitches imply that cognitive resources are usually focused on an outgoing event but distraction, inattention, or fatigue can lead to memory and perception enfolding in on one-another. In support of this Hughlings-Jackson (1888) suggested that people have two varieties of consciousness – normal (which process information from the outside world) and parasitic (or introverted – which processes thoughts and reflections of an inner mental world) [254]. When the activity of 'normal – extroverted consciousness is lessened by distraction, fatigue, temporary seizure, the evaluation of incoming sensory information depends more on internal consciousness, which derive from experiences already held, hence a new experience is 'misread' as an older experience – leading to déjà vu. A problem with this theory is that it is difficult to imagine how a function of consciousness could be 'parasitic' on other parts of consciousness. Either the metaphor is inappropriate or there are two forms of consciousness operating in parallel which may be combined in a perception of familiarity – straightforward perception of the environment (objective stimulation) and subjection reflection on internal experience (subjective stimulation. For déjà vu to occur the processing of these two functions is momentarily combined in a state of fatigue, causing an apprehension of temporal displacement.

Neurological Explanations

The basic premise of the neurological explanations of déjà vu is that it relates to brief and temporary neurological dysfunctioning characterised by a change or 'seizure' in the speed of neural transmission from specific receptor sites in the brain. Halgren (1978) proposed that it resulted not from a decrease but from an increase in the electrical outflow of the hippocampal gyrus (the area of the brain involved in encoding and retrieval) – this is qualitatively experienced as a 'misinterpretation' of familiarity [255]. Bancaud et al., (1994) proposed that the 'inappropriate feeling of familiarity' resulted from a nonspecific seizure of activity in the temporal lobe combined with current sensory input (the temporal lobe received information from both the visual and auditory cortices) so possibly either could be involved in the experience of déjà vu [256]. As Wild (2005) suggests, "the perceptual, mnemonic and affective regions of the lateral temporal cortex, hippocampus and amygdala" are implicated as regions of the brain that are activated in the occurrence of déjà vu [257]. This concept of a neural transmission delay from perceptual organs to the higher processing centres in the brain, result in a slight increase in the normal time it takes to transmit a message due to temporary synaptic dysfunction – a slowing in routine processing time of several milliseconds – the experience of this is misinterpreted as what is actually new information experienced as old information (see Grasset, 1904) [258]. It is possible that a fatigued state underlies this slowing in neural processing time, which temporary elongates the time between sensation and perception. A transmission delay involving two neural pathways rather than one seems to be the more cogent explanation.

Neural transmission delay is also the basis of another theory (Ephron, 1963) [259]; Humphrey, 1923 [260] which posits that the primary perceptual pathway goes to the dominant brain hemisphere while the secondary pathway routes through the dominant and the non-dominant. When delay from a non-dominant hemisphere is extended a déjà vu experience may result. Alternatively, an electrical excitation of one pathway in the dominant hemisphere may cause a temporal delay to be experienced in the secondary pathway – causing déjà vu.

Memory explanations may serve as the basis for theorising that implicit familiarity is the basis for déjà vu. If an individual processes information without paying conscious attention to it, subsequent processing may give rise to the sensation of objective familiarity in the absence of recollection. Déjà vu is contrasted from other memory responses through a strong impression of familiarity in absence of explicit recollection.

Memory Explanations

MacCurdy (1928) speculated that there are two components of nominal recognition response – affective reaction followed by familiarity (cf. Zajonc, 1980) – although two stages follow in quick and seamless succession – indistinguishable as separate processes, déjà vu results when initial affective stage is not succeeded by a clear cut memory match. Fleminger (1991) suggested that affective and cognitive channels of information processing usually work in concert but that déjà vu results from "aberrant activity in the pathway responsible for affective interpretation of percept's" [261]. Linn (1953) suggested that anxiety evoked by some aspect of the present situation disrupts normal functioning of a reticular activating system. Linn (1953) assumed that a change in arousal precipitates a déjà vu, rather than a specific affect associated with a stimulus [262].

Attention/Inattention Framework

A fourth explanational framework for déjà vu is that perceptual experience is divided into two separate perceptions as the result of distraction or inattention of perception. Déjà vu is caused when perception under diminished attention is followed by perception under full attention, the juxtaposition of these two experiences results in the diminished perception being attributed to a more distant past [263]. Leeds termed this a 'phenomenon split' and proposed that a physiological reaction such as an eye blink could divide these perceptions. Déjà vu has also been attributed to inattentional blindness [264]. When target stimulus is in periphery and the extraneous (ignored) stimulus is in the centre of the visual field (fovea) – inattentional blindness more likely than when target stimulus is in the fovea and extraneous stimulus is in the periphery [265]. Momentary distraction from a stimulus that is later perceived more clearly elicits déjà vu. Dixon (1971) suggested that during an initial brief stimulus exposure 'inhibition by interference' may cause parts of the stimulated field to interfere with the perception of other parts, a second glance of a scene after a first, results in initial disinhibituation which then matches present perception [266]. Déjà vu is difficult to re-create in laboratory or field experiments because it is almost impossible to stimulate the over-whelming sense of familiarity within a fleeting moment that needs to be concurrent with the realization that the stimulus could not have occurred in the context of the present.

THE PURPOSE OF DÉJÀ VU

The contribution which this chapter offers to the study of déjà is the contention that déjà vu, rather than being a feature of memory anomaly, is in fact a purposeful function of cognitive processing. Déjà vu is a useful experience, a product of evolutionary adaptation. While the experience of déjà vu may be one of heightened perception of brief temporal displacement it reorientates the experiencer to his or her subjective and objective present. This present is subjective in so far as the experience is qualitative and 'different' from the 'normal' experience of perceptual flow, it may even be a pleasant experience (the feeling of familiarity usually is), and is objective in so far as the experience is forced to perform a cognitive 'check' a 'real time' process of realisation that although the subjective experience seems familiar the circumstances and frame of the experience of the present environment are objectively different and could not have been experienced before. As O'Connor and Moulin (2010) suggest, "the overall evaluation of the déjà vu-eliciting situation sides with the higher-order metacognitive awareness of inappropriate recognition – the outcome is that the experient is able to function normally, does not modify his or her behaviour based on the errant sense of recognition, and can be left with a sense of wonderment at this insight into the normally concealed machinations of his or her mnemonic decision-making processes." [267] Déjà vu is then an aid to the framing narrative of consciousness, a device of memory that reorientates and familiarises the conscious state of the individual to the 'correct' perceptual flow of his or her current experience of environmental perception and experience. It might thus be a functional psychological adaptation to enable the growing mind to assimilate stimulating experiences into working memory.

Chapter 7

IRREGULAR PATTERNS

The natural world, including the functional composition of the human body is composed of complex systems. The human body and the human mind are complex systems. Consider the human lungs, for example. Have you ever noted the similarity in the pattern between the branches of a tree and the human lungs, both work to exchange carbon dioxide and oxygen, the one by photosynthesis and the other by molecular respiration?[11] Such patterns appear to be fractals that occur again and again in nature. Complexity and fractals are features of the natural world and the non-linearity created by complex systems. Consider snowfall, to take another example. Each snowflake is different, conforming to a catalogue of possible types, and each a fractal geometry as free-flowing water molecules are snap-frozen into tiny molecules of coalescing ice. This chapter explores the central ideas and principles behind systems complexity and argues towards an explanation for the puzzles of the human world, such as the iterative and generative qualities of language use, consciousness and evolution, based on hidden implicate order – fractals - and the *non-linearity* of emergent properties from complex systems.

Complex systems inform the natural environment at many levels, from the behaviour of water, of wind, the characteristics of meteorological conditions, climate change, and the transition between state changes, for example evaporation, water into steam and humidity into rain. Chaotic systems are said to be those that evolve with time and that exist in a state of confusion or

[11] Oxygen and carbon dioxide are exchanged in the alveoli in bronchial tubes. Capillaries surround the tiny air sacs or alveoli. When someone takes an indrawn breath oxygen flows from the alveoli to the bloodstream, and carbon dioxide moves from the bloodstream to capillaries into the alveoli. When they exhale carbon dioxide is removed from the lungs.

absence of discernible order but which nevertheless may exhibit some forms of spontaneous organisation when susceptible to perturbation [268]. Resulting chaotic forms of organisation or flow patterns on lower or higher dimensions occur in phases of chaotic attraction, either 'strange' or 'point' attraction in fractal dimensions. Examples of chaotic systems include the weather, turbulent water, epidemic infection among populations, the form of coast lines, and snowflakes. Chaotic systems are complex but not all complex systems are chaotic. Chaotic systems are aperiodic; they appear iterative but are in fact never repeating. And they depend on initial conditions.

DEFINITIONS OF COMPLEXITY

Complexity theory or CT is a term used to describe the theoretical and applied focus of systems thinking in an "environment of turbulence, flux, fragmentation, disequilibrium and uncertainty" [269]. For example, whether or not the universe has a mathematical structure or can be described by one, is a problem for complexity theory. CT is characterised by systems in which large networks of components with non-central control and simple rule of operation – produce complex behaviours, sophisticated information processing, and adaptation. As Lorenz (1972) proposed "[i]f the flap of a butterfly's wings can be instrumental in generating a tornado, it can equally well be instrumental in preventing a tornado" [270]. Hence further examples might be measurements conducted in different quantum systems occurring with different probabilities. Complexity theory is also characterised by non-trivial emergent and self-organizing behaviours, varying ontological states, of agency, of being, or of activity, and differing combinations of sets of relationships. Complexity theory is different from ambiguity, but may produce many outcomes, possibilities, from either groups of whole or fragmented subjects, nodes or network of connections, interacting parts, characterised by an unpredictability of emergent factors, parts, and behaviours.

Gilpin and Murphy (2008) state that complex systems are comprised of "numerous subsystems interacting with each other through multiple, nonlinear, recursive feedback loops [271]. They have highly organised multicomponent causal interactivity [272]. Simple systems in contrast may have interacting parts but they do so in linear and non-recursive ways [273]. Crutchfield and Machta (2011) argue that ". . . structural complexity arises from the dynamical interplay of tendencies to order and tendencies to randomness" [274]. Thus there is a tension in complexity between chaos and order, between systems in inflation and systems in entropy. Contrary to popular belief, chaotic does not necessarily mean crazy and non-understandable but rather a chaotic system is sensitive to change, to unpredictability and perturbations [275]. Complexity theory is also implicated in 'phase shifts' that mark the transition of one state to another in the natural world – a gas to a liquid, for example, or in cultural evolutionary terms, the amorphous structure and expression of human languages acknowledges an ongoing transition towards symbolic complexity in human culture. As Ekstig (2015) relates of the work of Laszlo (2009), a "universal flow towards ever-increasing complexity' is postulated, ". . . affirming the rate of evolutionary change to be accelerating at the socio-cultural level largely because the motors of mimetic change rely on information sharing (rather than energy/matter exchange) and therefore allows for faster rates of change than the biophysical level of genetic change" [276]. The complexity of language evolution is a process internal to human cultural development – a sub-product of consciousness and the neocortex enablement of symbol manipulation, it does not necessarily mirror the rate of physiological adaptation in the human species, though it may point to the adaptation and change of human subjectivities and epistemological complexity. Complexity may turn out to be a by-product of adaptation change in a system which has no 'directional bias' [277] other than a limit of 'minimal complexity.' As Ekstig (2015) points out, when an organism is faced with over-whelming environmental challenge it 'goes extinct' rather than evolving into less complex (and more robust because simplistic) forms [278]. Perhaps it is the relationship between gene expression and the environment that determines the rate of evolution, or the evolutionary potential of the species.

After periods of relatively rapid adaptation, species tend to remain unchanged for millions of years subsequently while the "new species normally stabilizes at the new niche in the complexity space" [279]. There is a relationship then, between the relative complexities of species according to the environmental niche it fulfils. However, the growth of complexity could be also related to the directionality of branching, with the complexity of species evolution reflecting the complexity of its evolutionary development also. Ekstig suggests that this might be a "deeper principle of complexification at work" positing, like Davies (2013) [280], that "not only will the most complex representative of the biosphere tend to become even more complex over time, but the median of the complexity distribution will also shift towards higher values over time" [281]. Thus as the Earth and its species grow older, the more the tree of life develops and adapts, the more complex it may become from its originary moment, or the evolution of the most basic life forms.

FACTORS OF A CHAOS SYSTEM

Chaotic systems have six main features. Firstly, they are sensitive to Initial Conditions, a property of the 'Butterfly Effect,' in which as Adewumi, et al., (2015) state, "[t]wo trajectories emerging from two different close-by initial conditions diverge exponentially from one another as the system evolves in phase space" [282], starting from similar co-ordinates the resulting pattern van be exponentially divergent but predictable if accurate measurements of initial conditions are known. Secondly chaotic systems are discrete. For any given time, interval there is only one future state that follows from it. They are non-linear in so much as inputs and outputs are disproportional one-to-another, the relationship between variables exhibits a dynamism as the system cannot be dismantled and assembled in the same way again. Fourthly, chaotic systems are instable, they have a 'sustainable irregular manner' [283]. They have attractors which are d-dimensional sets of states within the systems dynamics in which state that are near one-another asymptotically approach one-another. There are four kinds of attractors [284]. A point attractor if evolving to a single point, a limit cycle, characterised by predictable cyclic positions, a limit torus in which the systems trajectories are bounded within a ring torus – e.g., a rainbow; and a strange attractor, being a system of irregular shape that is aperiodic, never repeating. Sixthly chaotic systems have fractal dimensionality, with complex non-integer dimensions, repeating at smaller dimensions. The Mandelbrot set is an example of chaotic fractals (See Figure 5).

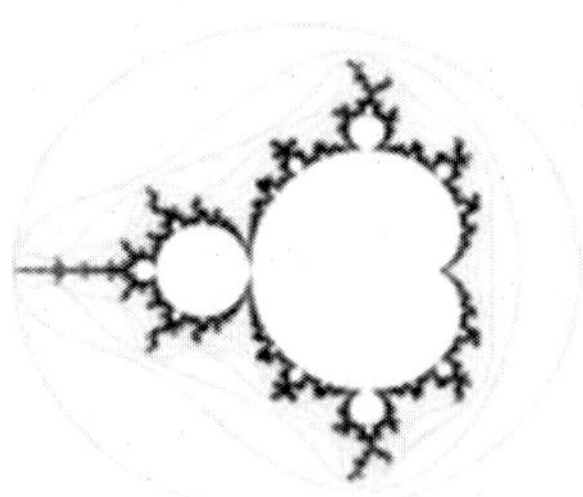

Figure 5. Example of a Mandelbrot Fractal.

However, complexity theory is also a system of thought that is applied to the quantum world. In the quantum world the observations of physical conditions may be influenced very subtly by the observer him or herself. Consequently, complexity theory is also accepting of the anthropogenic view.

As such complexity theory is not entirely an intrinsic property of a system, as Brachthauser puts it, "[h]ow we describe a (complex) phenomenon shapes and frames the methods and modes of explanation that we apply in our investigation" [285]. It stands to reason that complex environments can only be understood in thinking about them in complex ways with a variety of methodologies and frameworks.

According to Johnson (2012), complexity science is the "study of the phenomena which emerge from a collection of interacting objects" [286], of, for example, a crowd, or the behaviour of a virus through intercellular communication and signalling. Proponents of complexity theory argue towards an isomorphic (or two identical in form shape or structure) model of the world, that is described by "non-linear relations, multiple and contingent causation, and emergent properties/holism" [287]. Complexity theory thus tries to understand and provide structural order for unpredictable things, a set of finite but unknown alternatives described as 'determined chaos' (2008, p. 26) [288]. Chaos may be described by five factors. It is "science not state, is randomness emerged out of complexity, is mathematical term to explain an unpredictable behaviour which is not random, and is unpredictable behaviour in a nonlinear system" [289]. Thus chaos may share some characteristics of both entropy and 'phase shift,' complex processes are capable of becoming chaotic but nevertheless exhibit some patterns of order. As Briggs and Peat (1989) relate [290]:

> . . . the dissipative structures that led to life on the planet probably began at the chaotic interface between solid, liquid, and gaseous surfaces where there is a high flux of energy. Some scientists speculate that at this chaotic nexus, autocatalytic chemical structures like the Belousov-Zhabotinsky reaction constituted a form of protolife and that on early earth many variations of such reactions flourished. Responding to the far-from-equilibrium environment, the descendants of these first autocatalytic, self-referential, self-similar structures linked together to form a larger structure of feedback loops called a hyper cycle. One hyper cycle structure was RNA.

RNA (ribonucleic acid) and DNA (deoxyribonucleic acid) are nucleic acids that have complementary roles within living cells passing genetic information through a process called transcription. The cell creates proteins by using RNA through a process of translation. A cells chromosome is made of double-stranded DNA, which has sections of genes which code certain traits of an organism. The genes tell the cell to make proteins, using the cell's genetic code; DNA and RNA make the proteins which the gene codes for. A highly complex process.

Sanger and Giddings (2012) argue that a "complex system consists of numerous subsystems interacting with each other through multiple, nonlinear, recursive feedback loops" [291]. Determined chaos, represented shifts to non-linear conceptualisations of behaviour and change as possible descriptive analyses of complex physical and social phenomena. Chaos/Complexity theory may be characterised by [292]:

- Dynamism
- A-periodicy and instability
- Complexity with simple causes
- Non-linear, sensitivity to initial conditions
- Determinism – non-randomness
- Instability, a-periodicity, sensitivity to initial conditions
- Iteration/feedback

Furthermore, a complex system has the following features [293]:

- Similar but independent elements
- Constant movement and response to other agents
- Adaptiveness to ensure survival
- Self-organisation in which a system forms spontaneously
- Local rules which apply to each agent
- Progression to make the system sophisticated and larger
- Unpredictability
- Subtractivity

ANTI-REDUCTIONIST

As Bousquet and Curtis (2011) suggests, complexity is informed by systemic properties and relationships. Emergent properties arise through interactions over time, the 'whole is greater than the sum of the parts' and a dissection of a system into components may hamper the organisation of the system [294]. A relationship of reciprocity may arise between local and global levels of a system; examples may be the use of non-linear equations to describe phenomena which give rise to unknown recognizable patterns – turbulence of fluids or movement of plate tectonics [295]. As Valle (2000) also suggests, deterrence in complex systems is nonlinear, if deterrence is used for a particular event or effect, some else may happen in its place [296]. Thus, action and non-action may also both have causal agency in complex systems. It is possible that three some reductionist techniques could be used to identify potential pathways of cause and effect or system behaviour, however, ads Brachthauser (2011) suggests, "when faced with conditions of deep uncertainty that impede reliable predictions, process patterns can produce candidate explanations that limit the number and range of possible pathways based on plausible scenarios for specific cases derived from process tracing and/or computational modelling" [297]. In complex systems it is unlikely that any reductionist methodology could eliminate uncertainty completely in the classical sense, simply because complex systems are in themselves, not transparent to reductionist tools. Aside from the rather oblique observation that systems may act or act together in more complex ways than has hitherto been thought in mainstream scientific thinking and furthermore, that reductionist theories of science are self-confirming to a point, there does' however, exist a point at which reductionist theories of science fail, and complex (negative, for example) systems may take over that are non-explainable in reductionist ways – call this point the complexity recognition indicator. This can be illustrated in the tendency of organic life to evolve into more complex forms. As Ekstig (2015) states: ". . . [t]he suggested model implies that complexity is cumulatively increasing, giving evolution a direction, an arrow of time, thus also implying that the latest emerging species will be the one with the highest level of complexity.

Since the human species is the last species evolved in the evolutionary process seen at large, this means we are the species with the highest complexity" [298]. We simply don't yet have the necessary knowledge to fully understand our own species. As Mossman (2014) states, "[l]iving systems . . . reflect deterministic chaos—simultaneously expressing deterministic and chaotic behaviours" [299]. It may also turn out to be the case that different organisms in the tree of life evolve at different rates, for each species there is an evolutionary rate of adaptation in response to environmental constants in which the genetic potential of the organism in relationship to the environment may be realised. This could explain why the oldest organisms of the ecosystem, having been exposed to natural selection for the longest time, are some of the most primitive. It may even turn out to be the case that the evolutionary rates for different organisms are correlated in the 'tree of life' that given the existence of organisms V, W, and X the evolutionary development of organism's Y and Z may be inferred.

Computational Complexity

Ekstig (2015) maintains that 'complexity' is defined by "the number or parts or the amount of differentiation among parts within an individual" [300]. Despite the fact that as Sanger and Giddings (2012) claim, "there is no agreed on Metric for measuring complexity" [301], computational complexity is the term used to describe resources required in calculation to solve a problem which include such factors as time and space, complexity of instances, and complexity of algorithms [302]. Relatedly, the theory of algorithms computational complexity (ACC) measures time and space "to determine relations between the size of algorithms or machines and their efficiency" [303]. Within computational complexity, there are five classes of complexity: P, NP, NP-hard and NP-complete. Examples of problems include: Firstly, design of networks which have a minimum of routes costs; secondly, storage and recovery which have problems to maximise weights (products) and partitions (storage spaces) to obtain savings in storage; and thirdly, scheduling and allocation of priorities (problems of manufacture processes) [304]. Common classes of complexity algorithms are

- Polynomial time algorithms (constant time, linear time, quadratic time, cubic time, polynomial time, strongly polynomial time, weakly polynomial time, super-polynomial time and quasi-polynomial time)
- Sub-linear algorithms (logarithmic time, log-logarithmic time, and poly-logarithmic time)
- Super-polynomial time algorithms (sub-exponential time, exponential time, factorial time, and double exponential time) [305].

Furthermore, Ruiz-Vanoye et al., (2010) propose that a methodology to derive a metrics for complexity involves how much time an algorithm in a complex state requires to solve real problems and how much time algorithms with larger input require to solve real problems given a complexity of instances. Orponen (1994) derived a formula for measuring complexity of the instances of a decision problem: "the complexity of string x with respect to a set A and time bound t (ic1(x:A)), is the size of the smallest special-case program for A that runs in time t, decides x correctly and makes no mistakes on other strings" [306]. The complexity of the analysis should match the complexity of the problem analysed.

LIMITED DETERMINISM AND COMPLEXITY THEORY

Limited determinism refers to unpredictable, random, emergence behaviour. Although a system of behaviour appears random, influences acting on system, delimit behavioural outcomes. The Feigenbaum fractal (see Figure 6) is an example of strange attractor that appears in nature. The result may be 'randomness' or an inability to work out influences affecting system. A behavioural permutation, constrained by pre-determined influences, defends the validity of general predictions about system behaviour. Limited determinism describes the assumption that "apparent randomness in system behaviour is not random at all, but is instead . . . product of a vast number of factors that permit only broad, time limited predictions of behaviour" [307]. Complexity theory maintains a concept of equilibrium, which maintains a balance of system behaviour, characterised by constant adjustment to either attractors or repellers in a behavioural space [308]. These attractors and repellers are spaces within which a system functions. Perhaps unsurprisingly, attractors exert 'pull' on a system in terms of behavioural components and repellers have instability, pointing towards a movement away from the system – 'pushing' [309]. Complex systems evolve into linear states without environmental interaction, whether or not they engage in a complex system is dependent on the limited determinism provided by internal or external factors interacting within a system. Bracthauser (2011) has also suggested that there may be limitations to the various kinds of statistical modelling that are currently available to analyse complex environments – or at least there is necessarily some ambiguity in using current and emergent models of statistical analysis [310]. The purpose of statistical modelling is to detect patterns that may not have previously been noticed, hence the visibility or otherwise of a pattern could mean that causal explanations are missed, or even that some effects are not noticed. Furthermore, statistical pattern recognition is not necessarily explanation in itself; it may be an effect without a known cause. Systematic relations may exist within complex systems that are not reducible to statistical units and thereby constituent connective relationships that are not quantifiable, and in which organising principles may not be immediately apparent [311]. New methods of observation as well as combinatorial methodologies may be needed to explain the interaction of complex systems.

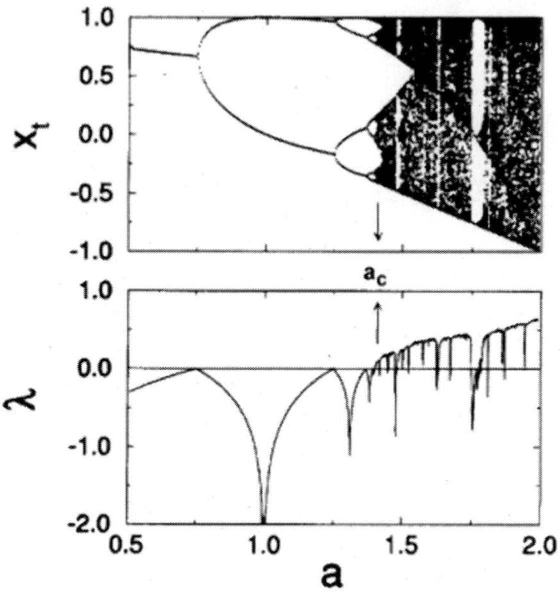

Figure 6. The Feigenbaum Fractal.

Consequently, statistical analysis as mode of explanation for complex phenomena is ambiguous. It is particularly important in detecting patterns that would otherwise go unnoticed. Missing the pattern would mean missing the explanation. Recognising the pattern, however, does not automatically constitute an explanation. Statistics deals with the problem of large numbers essentially by eliminating complexity and deliberately treating the individual elements as if they were not systematically connected. Analysis thus typically proceeds on the assumption that no information is required about how the elements are related and their relative position within the structure, ignoring any organising principle. Furthermore, it is also unlikely that mechanical explanations could fully come to grips with complex processes. As Brachthauser (2011) suggests, "[t]ypically, social and political processes are too complex to be captured by a single mechanism. Rather we find a nested hierarchy of mechanisms that operate in complex concatenated patterns. Often a mechanism is a component of a yet higher-level mechanism that imposes constraints on the behaviour of a lower-level mechanism. A mechanistic account may thus give rise to a cascade of levels." [312]. Cascading levels of system complexity may add dimensional robustness to understanding complex processes but just as complex events are not universally observable (cannot be viewed from all positions), the explanatory capability of such accounts will also be of limited use in describing cause and effect of complex systems.

COMPLEXITY THEORY AS A MODEL OF REAL WORLD PROCESSES

According to Bryne (1998) there are four appealing features of naturally occurring processes as description of complexity theory. These are: Firstly, treating nature as ontologically open and grounded in historical progression; secondly, viewing nature as hierchically structured with complex inter-relations; thirdly, regarding it as indeterminate and irreducible but nevertheless explainable; and fourthly, capable of forms of self-organisation devoid of psychological bias such as anthropomorphism [313]. Complex system behaviour may involve forms of transactionalism and tending towards equilibrium, if either attractors or repellers dominate, the system may be reorganised by a phase shift into another pattern of complex but non-random behaviour [314]. Thus complex systems may demonstrate forms of chaotic behaviour, they are indeterminant, with many possible outcomes given any set of inputs. Such systems are nonlinear, and bifurcate in 'choosing' a particular state at the moment of 'phase change,' (e.g., gas or liquid) [315]. Brachthauser (2011) also identifies a further features of complex systems, their non-decomposability [316]. Such non-decomposable systems are characterised by interdependence among components that is intense such that independent activity is compromised during its function. The organisation of the system is determined by a connectivity between parts that may be unexplainable (given current parameters – statistical or qualitative) both micro and macro processes react to produce emergent properties that are not explainable according to their constituent parts.

Sanger and Giddings (2012) argue that complexity is defined by at least four factors. These are: Firstly, that some aspects of complexity theory are self-confirming, possibly anthropogenic (observed within the eye of the beholder) [317]; secondly that complex systems are intertwined (parts are connected together in multiple, recursive, nonlinear feedback loops); and thirdly, that they are sensitive to initial conditions and hence may be unpredictable or have long-range effects [318].

COMPLEXITY THEORY AND EMERGENT PHENOMENA

In view of the fact that "complex systems are characterised by a multitude of convoluted interactions that collectively determine the system's dynamics. The possibility of multiple realizations, the context dependency of outcomes, and non-linearity between causes and effects create major obstacles for scientific inquiry," complexity theory is also concerned with continuity and change [319]. The concept of emergence argues that given a degree of complexity in an environment, new properties or behaviours may emerge that are not contained in the constituent elements. Concepts emerge from critical mass, path dependence, inertial momentum [320]. Uncertainly about initial conditions may lead to fluctuations in the behaviour of a particular phenomenon. Like chaos theory complexity theory is concerned with 'wholes' – that is with larger systems or environments among constituent elements, rather than with reductionist 'essences' [321]. Factors involved in complexity systems analysis are: multi-scale hierarchical organisation, emergent patterning, agent-based modelling, dynamical attractors and repellers, information flows and constraints, system-environment interaction, developmental trajectories, selectional ratchets, fitness landscapes, interaction across time-scales, and forms of self-organisation including critical mass, phase transition and autocatalysis and the butterfly effect [322, 323]. Hence complexity theories argue that causation is too multi-dimensional fast and unpredictable to focus the attention, shift from focus on causes to a focus on effects [324]. Mason (2008) argues that, multi-factor causality, which occurs through time and the absence of a generative force, results in 'emergence' and requires consideration of concept of 'structure' [325]. Consequently, complex systems may be characterised by the following features [326]:

- Output complexity is beyond the input complexity
- Adaptations within sub-systems may occur according to different levels of input and morphogenesis
- Strange attractors produce continual changes and heteroglossia within systems
- Stability and instability combine to produce variability
- Variability masks causes and effects

Complex nonlinear systems have behaviour which responds to small changes. As Woesler (2007) suggests, the complexity theory approach leads towards the possibility that physics, time and space "are discrete and that parameters of physics, which may vary in space and time according to simple laws, may be encoded in a finite number of bits" [327]. Thus complexity theory may help to explain such deeply embedded emergent phenomena as the relationships between electromagnetic particles and the symmetry of physical form. Complexity theories are thus sought to explain the hidden regularities of the world and the rules for computing measurements of quantum physics that contribute to the ontology of matter. Complexity theory is thus related to the possibility for a unified theory of physics in which probabilities might emerge automatically considering an infinite number of mathematical structures -- the logicality of possible worlds and relative occurrences [328].

Chapter 8

THINKING STRAIGHT

The human brain is the most complex organ known to be in existence, and it requires a lot of the human body's energy to maintain. As Harare (2014) states: "[i]n Homo Sapiens, the brain accounts for about 2-3 per cent of total body weight, but it consumes 25 per cent of the body's energy when the body is at rest. By comparison, the brains of other apes require only 8 per cent of rest-time energy" [329]. Thus the human brain is designed to be efficient in terms of processing capacity for its mass. Complex problem solving takes a lot of energy, so the brain invents heuristics or rules of thumb to save energy. Sometimes the body tries to save too much energy and cognitive heuristics can fail.

This chapter is about some common ways in which cognitive heuristics are made more complex by known error patterns. Cognitive heuristics involve quick 'rules of thumb' which help us to interpret the world around us, to process signs and symbols within the semiosphere, to filter out what is not relevant and perceive, understand and react to what is relevant. They probably arose from relation to emotion and cognitive faculties of the brain involved in detecting threats for the fight or flight response but also in some forms of relationships and exchange systems. Cognitive biases result from attempts to simplify the process of making sense of the environment, they are mental short-cuts, which while sometimes accurate can also be error-laden. As Stevenson, Cronin, Sur and Kording (2010) put it, "[n]eurons in the sensory system exhibit changes in excitability that unfold over many scales. These fluctuations produce noise and could potentially lead to perceptual errors. However, to prevent such errors, postsynaptic neurons and synapses can adapt and counteract changes in the excitability of presynaptic neurons" [330].

Cognitive biases are concerned with emotion, cognition, memory attribution, intention and attention; whereas fallacies are concerned with errors in the reasoning of argument – the two are related but not the same.

In popular psychology, here are six basic categories of bias that affect human reasoning. The heuristics of mental 'rules of thumb' people rely on to make 'judgements' or intuitions about the environment or not necessarily made from mental limitations but rather from intimations about the structure of the environment in which an organism is situated. The organism is likely to have a repertoire of strategies 'learned' in a storehouse of remembered experiences, which may be applies to a given situation. But when these rules of thumb are adopted unquestioningly, biases may occur. Mental biases can be grouped. The first of these is so-called 'biases of inattention.' These describe consequences of thinking error which stem from an inability, temporary or permanent, to concentrate. The second kind of popular biases are initial perceptions whereby information considered to be first-hand disproportionately effects future assumptions, calculations and judgments. For example, 'confirmation bias is the tendency to selectively search for information that confirms a hypothesis or preconceived viewpoint. The third kind of bias is that which arises from group participation. They occur as the result of common human reasoning errors, made when people think in groups. The fourth kind of bias category concerns biases that arise from the act of human agency. These may be conceived of as arising as a consequence of some fact of human embodiment. For example, the 'fundamental attribution error' is the tendency to overemphasize personal considerations and underestimate situational explanations for other peoples' behaviours. A fifth kind of bias category is that of 'framing effects.' This is characterised by the way in which different kinds of information may be perceived in the light of other kinds of information. A sixth category of bias in human cognition and rhetorical argument is that called biases of reasoning. This category involves the kinds of fallacy that arise from errors in argumentation logic. While the categories and descriptions of these major biases of human reasoning in popular psychology are by no means exhaustive, it is hoped that the following discussion will in some cases provide a form of ethical backdrop to the proposition that human reasoning, rhetoric and argumentation can and should be a rational process and that there are categories of distortion of human thinking that are commonplace and correctable.

Biases of Inattention

The first category of bias is one which is deceptively common. This bias might be applied both in everyday life and also in more formal situations, such as the conditions of scientific experiment. This category is termed 'biases of inattention.' There are generally thought to be four kinds. The first of these is 'expectancy bias.' 'Expectancy bias' is characterised by 'seeing what you want to see.' For example, everyone has probably experienced a situation where they thought they saw something – a lamp, a vase of flowers – because of familiar exposure, or because the shape or form or word is on their minds, and the existence of the object is recalled, even in its absence. A simpler example is looking up a phone number and substituting an eight for a three because you have just been thinking about units of eight.

The second kind of bias is 'selection bias': the inadvertent non-randomness resulting from an attempt to get a random sample. For example, you have just come from a meeting where you have been talking to a woman with red hair. You now need to select a group of five participants from a larger group of twenty for an experiment which requires a sample population selected on a random basis. The first person you see has red hair, so you select her first.

A third kind of inattention bias is the 'observer bias.' This is when the act of observation changes the results. The most obvious example is the rugby referee who gets in the way of the ball; but this kind of result has also been confirmed in quantum physics. The states of a quantum particle can apparently be affected by the position of the observer.

A fourth kind of bias that can be considered to be one of inattention is that of 'base-rate neglect.' This kind of bias is probably more likely to affect you when you have just woken up, or when you have low blood sugar and are very tired. It is characterised by a failure to define options, considering only what comes to mind – 'what you see is all there is.' Sometimes, though, the 'base-rate-neglect' can be useful – in preventing over-complication, for example.

Biases of Initial Perception

A second category of psychological bias is bias related to 'initial perception.' The first of these is termed 'anchoring.' For example, thinking of one number affects the value of the subsequent estimate, even on an unrelated issue. The anchoring effect may be simple (for example, the expected base rate of consumer prices when one visits a supermarket) but it may also be more complex (for example, the disparity in salary rates between shop assistants and professional workers may be in the order of tens of thousands or hundreds of thousands or millions, and each group of workers may have vastly different ideas about the value of each other's work).

A second example of a psychological bias of 'initial perception' is that of 'emerging preferences.' Once a decision is made about one alternative, a person is, on average, more likely to change his or her preferences and view additional information in a way which confirms the earlier decision. An example here may be a decision made on the basis of aesthetics or, perhaps, the support of a friend who has been criticised for a hitherto unrecognised behaviour.

A further example of a bias of 'initial perception' is termed the 'focusing illusion.' This illusion is confirmed by the fact that sometimes nothing in life seems as important as the issue that is within your immediate focus. A fourth example is termed the 'availability heuristic.' This is characterised as the process of making a judgment about frequency, based on the simplistic reasoning of the ease with which it comes to mind. This category can be deceptive – some elegant solutions to problems can take a lot of work, and repetition through familiarity can be drudgery. Common examples are when one question is substituted for another, you aim to report a category or a frequency of an event but you report an impression of the ease with which it comes to mind; or, asked to think of an animal, you think of an elephant because it is the largest of its class. A related category of initial perception bias is the 'statistical base rate.' When specific information about an example is evident, the facts about a generalised population to which a case belongs are sometimes underweighted or neglected altogether. Sometimes this is a good thing (in overcoming negative stereotypes about people, for example), sometimes it is less good, and (a particular job candidate is favoured because he wears a similar pair of jeans to those worn by the boss).

The 'halo or horns' effect is a sixth category of initial perception bias. This bias is evident in terms of forming first impressions of people or of the arguments they make. If people see one attribute that predisposes them to favour a person, they are slightly more likely to view that person favourably henceforth. Or more specifically it refers to a tendency to think that a person's positive effects, extend from one area of their personality into all areas. Or accordingly some negative features of a person's personality to extend to all features – according to the 'horns' effect. Similarly, if one outcome is chosen as the desired option, people are more likely to interpret subsequent information in a way that supports their conclusion, despite any evidence to the contrary. For example, people may be more predisposed to buy a product of inferior quality if it is in a colour that they prefer.

Biases of Group Participation

A third category of popular psychological biases are that involving group participation. A major example is In-group bias which people have a tendency to give preferential treatment to people they perceive to be of the same group. An example of this category is the 'bandwagon bias,' in which collective selection tends to be iterative. For example, one job candidate is preferred because of a groundswell of peripheral opinion. Second examples of biases of group participation are 'casual base rates' – generalised statistics may be incorporated as the norms of an individual case and combined with other case-specific information. For example, your doctor diagnoses you as having high blood pressure and also presumes that state is due to work stress and also infers that your diet is cholesterol rich. A third kind of fallacy is given the Latin name '*Argumentum ad populum*' fallacy. This refers to reasoning that suggests that because many people favour an idea, you, or others, should too. This kind of bias is also related to the 'democratic fallacy,' which involves the fact that although the majority has a view, that in itself has no bearing on the truth or falsity of the view. Just because the majority believe X is true doesn't give us any evidence to allow us to conclude that X *is* true. A fifth kind of bias that follows from some forms of group participation is the '*consensus ad idem*' (for example, the 'meeting of minds' that is said to exist when a contract is reached).

Anthropomorphic Biases – Those Fallacies which use Human Agency

The fourth category of biases can be attributed to aspects of being human. They may be concerned with human embodiment or the way that the human body is conceived within processes of reasoning. The over-arching category here is that of 'anthropomorphism' – attributing human characteristics to inanimate objects. Relatedly, 'zoomorphism' is the tendency to attribute animalistic qualities to human or non-human entities. An example of one form of this reasoning is the 'straw man fallacy.' This type of argument is characterised by responding to a statement so as to deliberately distort it and weaken it. In rational discourse one should address the argument and not the person behind the argument. Thus the statement, 'The reason I don't like politics is because I don't like politicians' is a 'straw-man' fallacy, as well as being an over-generalisation. The '*argumentum ad verecundiam*' fallacy is the reliance on a particular belief due to the testimony of a person who is not an authority on it. It would be like a priest preaching about science. We commit the 'narrative fallacy' if flawed stories about our past shape our views about the world and our future expectations. You may have a schema of 'shyness' instilled in you and minimise all contact with others just because you recall the misnomer as a traumatic event. Furthermore, a common anthropomorphic fallacy is over-confidence, whereby the individual tends to discount the quality or quality of the evidence. Another way of putting this is that the confidence that people have about their beliefs depends more on the quality of what they can see and hence tell themselves, even in frames of reference in which they can see not much at all. Finally, the 'hindsight' bias occurs when someone tells you that they 'knew it all along,' and is usually applied after the fact.

A sub-category of biases of inattention are the psychological biases that may stem from inauthentic experiences of self. There are generally thought to be seven kinds of psychological biases stemming from mis-comprehension of self. These are: First, 'denial' – a refusal to admit that *x* or *y* is occurring; second, 'rationalisation' – a tendency to ascribe some unpleasant perception to a cause outside the self; third, 'confabulation' – a tendency to invent concepts that protect self-image; fourth, 'reaction formation' – the tendency to assert the opposite of what you might otherwise know to be true about the self; fifth, 'projection' – ascribing deficiencies to another; sixth, 'intellectualisation' – transforming an emotionally difficult fact into an intellectual problem; and seventh, 'repression' – the tendency to block the retrieval of painful memories [331].

'Attributional' biases are a further class of errors in perceptual deduction that follow from misaligned causal attributions, usually due to a lack of knowledge of the other. 'Personality bias' attempts to explain the behaviour of others according to their personality dispositions. In trying to a see a situation either from another person's point of view or in another way, we may see more situational causes [332]. 'Self-situational bias' is the over-attribution of our own behaviour to situational factors. Group membership also produces biases. *Out-group* members are seen more differently than *in-group* members.

For in-group members positive behaviour is attributed to personal dispositions and negative behaviour to situational factors; but out group membership is explained in the opposite manner. Attributional bias is also culturally produced – in individualistic cultures we may have a greater tendency to believe that individuals are responsible for their behaviour, whereas in collectivist cultures people are more aware of situational constraints [333].

BIASES OF FRAMING EFFECTS

A fifth category of bias is that of 'framing effects,' the way in which the order and sequence in which information is presented effects perceptions of the outcomes. One such bias is the 'Gambler's fallacy' in which there is a tendency to believe that future possibilities may be changed by past events, when they will be unchanged by them, or the chance of one set of occurrences happening, will be the same as it was. An extension of this fallacy is the 'hot-hand' bias which is the expectation of 'streaks' of similar events in sequences whose probabilities are independent [334]. A further bias of framing effects can arise from simple matters, such as the pricing of consumables, to the calculation of rates of interest on a loan. Whether an item is described as ninety per cent fat free or ten per cent fat invokes different responses, despite the fact that mathematically the proportion is the same. A second kind of 'framing effect' is the 'availability cascade.' This is similar to a form of mediated 'moral panic' and is described by a self-sustaining chain of events that may escalate from a relatively minor event to a larger-scale event – for example, some bullish or bearish stock-market runs may be triggered in such a cascading way. A related technique that shows a proportion between two measures is the 'correlation co-efficient,' which varies between zero and one and is a measure of the relative weight of factors the two measures share. Many mathematical probabilities are examples of correlation co-efficients. Another form of framing effect is the hindsight bias by which is meant that through distortion of memory, peoples' recalled judgments of the likelihood of the event are closer to the outcome than their original judgments were.

The 'illusion of validity' suggests that, frequently, general predictions are little better than random guesses but sometimes assuming validity beyond their statistical relevance. The 'planning fallacy' is also a bias of over-confidence. It suggests that plans and forecasts that are unrealistically close to 'best-case' scenario may be demotivating, have over-inflated positivity and could be improved by consulting statistics of similar cases. One way to ensure the validity of planning forecasts is to use a reference forecasting model. This has three main components. First, an appropriate reference class is identified; second, the relevant statistics of the reference class are obtained and used to generate a base-line prediction; third, use specific information about the case in question to adjust the baseline prediction (this is similar to the use of Bayesian[12] statistics, in which one measure of probability can be updated by new evidence). The 'evaluation framing effect' suggests that, when viewed against a neutral reference point 'adaptation level,' the principle of diminishing sensitivity applies to sensory dimensions and evaluations in changes in wealth and loss aversion; when comparisons are made, losses seem larger than gains. If I view a product purchase in terms of the deficit to my bank account rather than in terms of the good or service it may purchase, I may be less likely to buy it. Two related effects are the 'possibility effect' and the 'certainty effect.' The possibility effect is that highly unlikely outcomes are given disproportionately more weight than they deserve; and the 'certainty effect' is that outcomes are almost certainly given less weight than their probability justifies. A further maxim of over-estimating and over-weighting is that people tend to over-estimate the probability of unlikely events. Some metaphors people use to 'think by' are just plain convenient, and they emerge as extensions of activities that people engage in to functional effect, or from the frustration with these. The verb 'to crane' for example may refer to an explanation or solution that has emerged 'from the ground-up'; whereas a 'skyhook' serves as a metaphor to explain a solution imposed from a macro 'on high' view of a problem.

BIASES OF REASONING

A sixth category of psychological bias relates to more straightforward categories of 'rhetoric, reasoning and argumentation.' Denying the antecedent is a question of inference. "If A then B" does not necessarily imply "not A therefore not B." If I turn on the tap so the water runs does not imply that if I don't turn on the tap the water will not run. Or the opposite fallacy may be obtained by 'affirming the consequent': "A therefore B" leads to "B therefore A." That the sun is shining so I can see doesn't necessarily mean that I can see

[12] Bayesian statsitics is names after Thomas Bayes who lived from 1702-61 and is based on Bayes Theorum which is a formula of statistical influence in which probability is described based on conditions which relate to the probable event. P (A/B) = P(B/A) P (A) / P (B).

so the sun must be shining. A related term is 'equivocation.' This is a term that has more than one meaning. 'Don't settle for anything less' comes to mind. A fallacy occurs when words with multiple meanings are used for deception in argumentation. An example is M-P; S-M therefore S-P, or, M-P; S-Q therefore S-P. The '*reductio ad absurdum*' fallacy unfairly attacks an argument by extending it to such extreme lengths it looks ridiculous – for example, 'all morality is subjective.' Similarly, two wrongs don't make a right.

Rational arguments should almost never use force. An ideal argument allows people to see the truth on the basis of evidence. The only force that an honest arguer should need to use is the force of reason. Similarly, claims to authority may involve the uses and abuses of expertise. It is legitimate to appeal to the views of experts if they are relevant to what is being argued. However, 'tradition' in argumentation can be both used and abused. Established ways of doing things are commendable if they stand on their own merits and tradition can be seen as a series of precedents. However, the fact that things have always been done a certain way is not in itself a compelling reason to keep on doing things that way. It is relevant therefore to always consider more than one source in searching for solutions to a problem. In rhetorical argument the 'either-or-fallacy' is reasoning that sets up false alternatives: if inferior reasoning is rejected, then the other argument must be accepted. In comparison, the 'post-hoc' fallacy involves the mistaken assumption that one event is caused by another because they occur sequentially. The fallacy of expediency mistakenly assumes that efficiency alone is sufficient to determine the worth of an action. Another form of reasoning fallacy is that of a failure to reach a conclusion to an argument – human reasoning is purposeful and we usually discuss problems in order to resolve them. 'Agree to differ' is the classic response to this situation. Similarly, simplistic reasoning is argumentation that relies on an over-simplified solution to a problem. In some circumstances over-simplification may lead to distortion. 'Men are interested in only one thing' tends to limit their options somewhat.

Other forms of fallacious reasoning may involve the undistributed middle term. To make arguments logically coherent the distributed middle term, which appears in the premise and not in the conclusion, needs to be a universal term and distributed at least once between the premises of the argument in order to have the proper scope to make a connection between the major and minor term. 'Begging the question' suggests that the argument lacks real premises. The point that needs to be proved to be true is simply assumed to be true. Making a full assumption is assuming something is true when it is taken to be true without actually assuming it is. One form of test for this form of reasoning is whether the argument involves contradiction – is it patently absurd? 'All people are created equal; therefore, all people are treated equally' is one example. The quantifying of quality is always problematic. In one sense an argument can be made that if quality can be perfectly translated into quantity there would be no need for a distinction between the two terms. There are many aspects of the world that do not translate into quantitative terms: for example, colour perceptions or peoples' emotions. Similarly, a red-herring argument is one that diverts attention by containing emotionally volatile information that is calculated to agitate a specific audience.

An argument which is *a fortiori* observes that what is true in one set of circumstances is also true for other circumstances, for example, seen one, seen them all! *Reduction ad absurdum* involves taking the logic of a statement to an absurd conclusion, for example, you've got to roll with the punches, but I never expected to get flattened! An *innuendo* involves discounting the faith of an audience in a person by subtle repetition of a positive statement, for example, 'the people were happy but he was happiest!' *Anticategorica* involves reversing the argument of an opponent, for example, 'you say we're the 'underdog' but we always win from behind!' *Anthypophora* involves asking and answering a rhetorical question, for example, 'did you only drive all this way for one reason – I think not.' *Paromologia* is a commonly used form of argument that admits a weaker point then makes a stronger one, - I was hear five minutes already, don't keep me waiting for ever! [335].

Reductionist arguments are a form of over-simplification. While they may seek to reduce something to the 'bare essentials,' they may frequently go further and render something less than it might be than if it were treated as a whole. 'That's all there is to it' is the common rejoinder here. Misclassification is the premise that we might come to know something by associating it with another thing. The grouping of one thing with another does not necessarily imply that the two are meaningfully of the same category. The exception to this is the fact that chimpanzees share ninety-seven per cent of their genetic material with human beings.

Likewise, an inability to disprove an argument – that fact that there is no direct evidence against a position – doesn't constitute an argument in favour of a position. It is like the agnostic beliefs in God. The fallacy of the 'false dilemma' involves a situation involving several possibilities. The speaker attempts to persuade the listener that there are in fact only two. This dilemma is termed false because the actual field of alternatives in the argument is distorted. There is a related fallacy of causality – in the cause and effect relationship the cause is also prior to the effect. This requires that the cause be present before the effect can take place. The Latin term for this fallacy is *post hoc ergo propter hoc* which translates as 'after that, therefore on account of that.' This is an example of 'putting the cart before the horse.' A final category of argumentation fallacy is that of 'special pleading,' in which information is omitted because it might count against a position that is being promoted. For example, someone presents a wallet to the police-station to collect a reward, but neglects to say that they had taken the money from it before handing it in.

The purpose of this chapter has been to define, from a qualitative perspective, the many and varied accounts of cognitive bias that are known within popular psychology, to classify psychological concepts of bias and to make inferences about the kinds of thinking that may give rise as conceptual causes and to show examples of the biases from everyday life. The value in doing so is not obvious. First, identification of cognitive biases might give us insight into the kinds of fallibility that can enter into our thinking; second, recognition of cognitive biases can lead to the correction of unhelpful errors that may enter into our thinking; third, classifying psychological biases may help us to make inferences about the cognitive conditions that lead to biases; fourth, increasing our awareness of cognitive bias may make us less inclined to erratic thinking. It may even improve our self-understanding, understanding of others, our communication, our sense of self-motivation and our understanding of the reliability of our cognitive experiences.

Chapter 9

YOURS OR MINE?

Emotions and feelings are a form of 'social economy' that runs like a thread in our lives. Whether we feel pride over a friend's sports achievements, embarrassment because we forgot to keep an appointment, anger that we couldn't get to put our point of view across, or sadness because we have encountered one of life's lessons in futility, according to 'emotions as social information theory' (ESI) people use other people's emotional expressions to disambiguate everyday life. Emotions are a form of shared and internalised 'language' by means of which we enhance our experience of our lives and that of others. Despite the importance of emotions, this chapter is concerned with their influence on whether we compete or cooperate with others. We will consider two emotions in greater depth later.

Emotions are relevant the interpretation of situational construal that is necessary in assessing whether situations and cooperative or competitive. The assessment of whether situations are cooperative or competitive is a significant influence on human behaviour because it influences collaboration, trust and hostility, among other outcomes [336]. Emotional expressions play a large role in making sense of socially ambiguous situations. The two poles are observable from the perspective that, the emotional expression of happiness through smiling and laughter signals group-affiliation and pro-social intentions and enables the construction of social relationships. Anger, on the other hand, is a display of negative emotion and may be a precursor to hostility [337] Thus the human capacity to interpret and create meaning is in part based on the social behaviour of others, in which there is a bi-directional link between a situation and the perception and interpretation of others' emotions [338]. As Han (2013) relates:

> Agents make commitments towards others when they give up options in order to influence others in a certain way. Most commitments depend on some incentive that is necessary to ensure that an action (or even an intention) is in the agent's interest and thus will be carried out. The capacity for using commitment strategies effectively is so important that natural selection may have shaped specialised capacities to make this possible [339].

CROSS-FUNCTIONALITY

While cross-functional alliances may be beneficial for organisations in some contexts, they may also detract from some areas of performance, due to lower levels of shared trust that may arise between organisations [340]. As Luo, Slotegraaf, and Pan (2006) state, ". . . interfunctional rivalry decreases the level of trust in information, which suggests that cross-functional competition coupled with co-operative ability could decrease a firm's customer-based performance" [341]. By comparison if cross-functional cooperative intensity and competition is co-incidental it can increase both customer performance and financial performance [342]. Increases innovation and market learning, improves transfer of complex knowledge and adaptation. By comparison, people who see the world in negative terms, experience greater degrees of stress and anxiety when faced with adversity and this is more closely correlated with people who must win at all costs, as they believe that the world is hostile to their goal achievement. People who believe that intelligence is adaptive and can be acquired tend to adopt a mastery orientation. However, Franken and Brown, (1996) also posit that people with a strong need to win are less instrumental than those who like difficult tasks and to perform well [343]. The need to win is correlated with more performance qualities – modelling and appearance than mastery qualities (planning, and being open-minded and having a strong will to win). Performance orientated people are more concerned with pleasing others, whereas mastery types are interested in developing competence. It is also thought that those with a strong need to win are not as adaptive and tend to adopt patterns and denial and behavioural disengagement when faced with adversity [344].

Coping Strategies

People with a strong need to win, as compared with people who like to perform well or whom like difficult tasks tend to have poorer coping skills, to disengage behaviourally, to see the world as more hostile, to have a view of intelligence that is fixed, and to view the road to success as coping from appearance [345]. People with a need to win, or have a poorer self-esteem tend to have less hope than people who like to perform well or to do difficult tasks. As Franken and Brown (1996) argue, there is an increasing body of evidence that a competitive/win orientation tends to undermine student achievement and performance [346]. Perhaps this is because people who have increased satisfaction from improving performance even when not winning, or whom relish difficult tasks, have more resources and resilience than people who like winning for winning's sake.

Innovation

Companies encouraged working with strategic partners in increasingly competitive global markets in wake of financial crisis at the end of the first decade of the twenty-first century, but often alliances fail. True collaboration beyond formal governance structure is hard to achieve, however adaptation and knowledge sharing is frequently necessary for innovation. Zhang, Shu, Jiang, and Malter (2010) point out that knowledge acquisition is positively related to knowledge creation and that knowledge creation is positively related to innovative performance [347]. Furthermore, between firm or inter-firm cooperation in strategic alliances is positively related to a company's knowledge acquisition from its partner. Zhang et al., (2010) point out that, international alliances increased the effect of knowledge creation on innovative performance but do not increase the effect of knowledge acquisition on knowledge creation [348]. But knowledge redundancy greater in domestic alliances than international alliances. International alliances usually formed for the purposes of innovation. Competition can result in aggressive knowledge acquisition from partner firms but alliances may fail as a result of excess inter-company rivalry but learning and cooperation can be achieved when company's commit to working together.

Cooperation with Competitors

Coopetition originated in the 1990s (Brandenburger and Nalebuff) as a term denoting competition with competitors [349]. It has been referred to in slang terms as being akin to 'sleeping with the enemy.' Competition and co-operation are mutually promoted by each other, providing an organisation with an advantage. Although centralised and autonomous firms tend to be more competitive, coopetition denotes a business environment of having heterogeneity in resources, the closeness of an activity to a customer, and a condition of connectedness with a competitor [350]. The forces that drive cooperation are a low level of threat, or greed or fear, a high level of communication among partners, and a reciprocity principle of interaction in partners. If organisations don't identify with one another, they are more likely to compete. Cooperation may involve any of four kinds of horizontal relationships. These include – coexistence, cooperation, competition or co-opetition [351]. In interconnected but rival organisations, forces of cooperation and competition may influence each other producing closely related but separate parts. Related to cooperation between companies is 'market commonality' which is the degree to which the presence of a competitor occupies a market place that overlaps with another company, whereas 'resource similarity' is the extent to which a competitor possesses strategic resources akin to those of another firm [352]. Increasingly competition and cooperation not seen as polar opposites but rather as co-existing to different levels and extents, and manifesting in different relationships. The outcome of inter-firm cooperation may include any or all of perceived success, concurrent financial gain, cooperative relationship goal achievement, product or process improvement and resource sharing [353]. Furthermore, for Peng (2012) the degree of market overlap determines whether firms are direct or immediate competitors [354]. Competition may increase cooperation between rivals whereas cooperation increases scale of market power. A strategy of co-opetition which balances competition and co-operation may lead to better performance in two ways. Firstly, by permitting the attainment of new performance levels, secondly, by changing timeframe permitting goals to be realised earlier [355]. Forms of cooperation include co-procurement (minimising prices), co-marketing (wider reach), co-distribution (simultaneous market presence).

Cooperation/Competition Continuum

As Rubin (2014) points out, the fundamental economic unit is the transaction. A transaction is cooperative. Market economies are supposed to produce consumer surplus, derived from transactions which are co-operative. The competition in a market itself drives co-operation. Competition established efficient terms for co-operation [356]. Whereas competition is zero-sum thinking, co-operation is positive-sum thinking. Economic activities are designed to maximise utility winning or losing having the greatest share or the highest income is incidental to the goal. The consequence of a transaction is that all parties seek to benefit from it. Thus transactions involve co-operative people who engage in transactions might not think them as co-operative but markets are supposed to facilitate unintentional co-operation [357]. The production of goods and the exchange of goods are co-operative acts, motive may be competitive but actions are co-operative [358]. What does a predator do, produces a good (co-operative act) and sells them below cost, or purchases at higher prices to deny resources to a competitor? Sale of good co-operative act. In pure competition – monopoly – effectively no competition. Specialisation a form of co-operation, generate surplus through co-operative acts. Competition best possible method of ensuring optimum co-operation. Competition for right or ability to cooperate. Market is a transcendently space in which wee interact distantly with millions of people throughout the world, people from the past and future. For Rubin (2014), market transactions involve people from geographically separate locations who can cooperate, and the market transcends time and space [359]. Most market intervention aim to restore co-operation, and many market failures are due to problems with competition – such as cartelization or monopolisation of markets. Forces such as asymmetric information and co-ordination problems, hold-outs, free-riders can prevent proper functioning [360].

When to Compete or Cooperate

Argyrols and Schon (2010) argue that we do not consciously choose how to behave when we encounter unfamiliar or threatening situations but frequently we respond defensively, engaging in competitive as opposed to interactive behaviour [361]. Fair play is important, if people feel unfairly treated or in danger of being unfairly treated, or not at risk of a fair share, they tend to expend time and effort in frustrating another party whilst acquiring more of their own [362]. As the American President Franklin D. Roosevelt stated, “Competition has been shown to be useful up to a certain point and no further, but cooperation, which is the thing we must strive for today, begins where competition leaves off” [363]. Trust is an underlying foundation for success in the business world; trust is engendered through co-operative relationships, and weakened in threatening or competitive situations.

Mitigation

As Malhotra states, the ‘competitive arousal’ model argues that elements of the strategic environment such as competitiveness in rivalry and time pressure can coincide [364]. Competitive behaviour is defined as the pursuit of assets which are scarce, it is also mechanism for creating value. Competitive actions can lead to destructive behaviours. For Ku et al., (2005) desire to beat rival bidders can lead auction participants to pay more than an item is worth to them [365]. But desire to win not necessarily irrational, Messick and Thorngate (1967), people may pursue relative gains more vigorously than personal gains because they do not like receiving less than others [366]. Two basic conditions for competitive action – scarcity and contestation [367]. An underlying factor of motivation to maximise relative payoffs is a form of social comparison, or the tendency to judge the ‘goodness’ of an outcome by comparing it to others [368]. Sometimes this produces the reaction of motivation shifting to beating the other side.

Competitive arousal model, characteristics of competitive situation (rivalry and time pressure) stimulate physiological arousal, motivation moves away from goal of attainment toward beating competition at high cost. Pronounced competitive arousal can dampen aggression which might have costly personal consequences. many competitors at outset many contests (for example, auctions, corporate acquisitions, promotion contests) whittle down to head-to-head between parties. Visceral reactions to environmental stimuli at moment of decision are powerful motivators of behaviour.

Balance in Competition and Cooperation

Basic economic tenet that competition drives resource allocation towards efficient outcomes under neo-classical perfect competition [369]. Price is sharpest weapon, co-ordination of prices a form of competition. Cooperation might preclude multiple paths to invention and dull incentives to be creative [370]. The conventional view of competition is that it is antipathetical to some forms of business behaviour because involve agreements between competitors. Yet although activities of business are independent, nevertheless inter-related [371]. Profitable only if a competitive investment does not exceed limit set by demand, volume of investment reaches right level, so market economy is in a state of 'ubiquitous uncertainty' – the horizon of consumers is unlimited and companies to not know their competitors future actions, or state of technology, etc [372]. As Han (2013) states, "[k]nowing the intention of an agent is a means to predict what he will do next or might have done before. The recognizing agent can then plan in advance to take the best advantage of the prediction, or act to take remedial action" [373]. Uncertainty is highest is at its highest when new actions are being taken or new processes and or technologies being developed. As Jorde and Teece (1989) point out, exchange transactions are unilateral and not bilateral [374].

Organic and Mechanical Metaphors

Ims and Jakobsen (2006) argue that the mechanistic worldview is more closely allied with competition, and the organic view with cooperation [375]. Market behaviour involves economic, societal and environmental values – transparent co-operation within an organic world view involves intrinsic as well as instrumental values [376]. Competition natural state of affairs in hostile market, aggressive individuals motivated by self-interest. Need independence to asset power. Different in organisation, business not war and relations between companies are predominantly cooperative – balance between competitive and co-operative [377]. Many large firms have bilateral arrangements – value-adding strategic alliances, joint ventures [378]. On one pole of the continuum, a competitive game consists of atomistic players in zero-sum game, whereby co-operation is instrumental; and at another pole there is a collaborative agreement among the strong to restrain competition [379]. Both seem to imply the operation of a causal and deterministic universe in which options are limited (which may in fact correspond to many real world business options). On the one hand, "[t]he mechanistic paradigm pre-supposes a dualistic separation of 'nature' and 'life,' whereas in the organic concept of nature, 'life' and 'mind' are interwoven with matter and motion" [380]. In the latter, life exists for itself, has intrinsic value Capra (1982) [381]. Living systems have non-linear connectedness. Relatedly there is a view of a 'limitless' market which could not be reduced to mechanical parts, instead a market consists of interconnected partners integrated in a natural system but how realistic is this view in fact, however complex the market might seem? [382]. Cooperation allows more integrated solutions than atomistic and competitive economy. The organic metaphor of the tension between co-operation and competition is well illustrated by a quote from John Donne[13] (1624):

> No man is an island, entire of itself; every man is a piece of the continent, a part of the main. If a clod be washed away by the sea, Europe is the less, as well as if a promontory were, as well as if a Manor of thy friend's or of thine own were: any man's death diminishes me, because I am involved in mankind, and therefore never send to know for whom the bell tolls; it tolls for thee [383].

Cooperation

Cooperation can be seen as a kind of social learning. Agents acquire knowledge from other agents by 'being told' as well as learning from experience, social learning is a useful way of increasing abilities and common human practice. As Han (2013) relates, ". . . in social learning, agents in a population can observe the behaviour of others and the outcomes of that behaviour. They may copy the behaviour of others whenever these appear to be more successful. The accumulated payoff from all interactions emulates the agents' fitness or social success and the most successful agents tend to be imitated by others" [384].

Positive feedback is a type of co-operation. Feedback can be given on outcome, intrapersonal factors, and/or actual performance. Similarly, there may be three types of outcomes (task enjoyment, positive effect and perceived competence) [385] who receive positive feedback report higher levels of competence and positive effect.

As Shaomeng (2011) defines it, in the context of the history of civilisation, Greek democratic theories represent cooperation, while Roman Republicanism represents competition [386]. Shaomeng (2011) argues that

[13] John Donne (1572-1631) was a English poet in the metaphysical genre. His style is that of vibrant sensuality and metaphorical invention. A cleric of the Church of England, in 1621 he became the Dean of St Paul's Cathedral, London. The quotation appears in *Devotions upon emergent occasions and several steps in my sickness - Meditation XVII*, 1624.

cooperation requires that all parties participating in cooperation accept the same rules of cooperation, which can be carried out only when all parties consent, to expected goals, labour, procedures and resources [387]. John Rawls defines social cooperation as how the basic structure of social institutions should be arranged such that all members of society would accept the distribution of benefits and burdens of co-operation [388]. Justice of fairness. In competition model, democracy doesn't need exact place to deposit sovereignty, no single office represents people, no single mechanism ensures legitimacy. Neutrality lies in balance between government. Legislature and people [389]. Liberty principle asks representatives to view society as system of cooperation, difference principle asks to take society as system of cooperation [390]. Competition model recognises social inequality as natural result of peaceful competition, people given sufficient leverage to liberty principle to counter-balance effect of social inequality [391].

As Osarenkhoe (2010) observes, inert-firm dynamics involves both cooperation and competition – coopetition, enhances internal resources and market shares of actors [392] Social exchange theory considers relations a dynamic process – used as a framework to understand buyer-seller and buyer-buyer relationships [393]. Competition dynamic situation occurs when several actors in specific area for scarce resources and/or produce and market have very similar outcomes. In long run competitive behaviour can lead to monopolistic position of firm within an industry instead of perfect completion [394]. Successful company able to create, invent and innovate [395]. However, Cangello argues that human beings cooperate more than expected in terms of utility, people may have developed an instinct towards cooperation, and hence a propensity to learn cooperative behaviours [396]. Firms cooperate for purpose of achieving a common goal, share resources with competing actors, for purposes of learning or sharing organisation expertise [397]. For business type sin competition model. The monoplayer – (low competition; low cooperation) – organisation does not interact significantly with competitors, low degree of competition and cooperation [398]. A contender (high completion, high cooperation) an organization that views competitors for market power, competitive position and market share – a high degree of competition and cooperation. A partner has low competition and high cooperation, shared synergies – complimentary resources and capabilities [399]. An adapter has high cooperation and high competition – organisations mutually depend on one another to achieve respective goals [400]. A Nash equilibrium arises (from game-theory) when a player has chosen a strategy and no player can benefit by changing strategies while the other players keep theirs unchanged" a form of co-opetition.

Chapter 10

LICENSED DIALOGUE

There are many different types of human expression – body language, sign language, haptics (another word for touch sensation), quiet conversation, free-flowing dialogue, structured dialogue, transaction, interviewing, and so on. What we do when we converse and exchange conversation and dialogue is exchange information, about ourselves, our relationships one-to-another and about the state of the worlds that we know. One of the most significant places we talk in in our lifetimes is the workplace. We converse with our colleagues about work, our environment, and wider politics while some forms of conversation take on more structure because they entail decisions about human resources. Interviews in the workplace are about a variety of things, assessing competencies for job tasks, assessing aspects of personality for collegiality and resilience, and for exchanging information. This chapter looks at some of the psychological concepts in interviewing to make inferences about the kinds of complex inter-personal exchanges that they represent.

An interview is like a window on a person's mind. Opening the window, we can see what lead to the information being exchanged. Aside from haptics and body language, dialogue is the best chance we have of gaining insights of other people, and of knowing them, aside from pure observation. An interview in the workplace or professional contexts, range from conversational exchanges on any number of topic contexts, to a structured competency dialogue for employment prospects purposes, to a clinical interview given by a physician to elicit information about a physiological condition or state of health; or even a police witness interview and intelligence interview, which attempts to gain information about a topic best left out of public view. Central to each is making sure that the interviewee clearly knows the purpose of the interview and has the competency and aids necessary to do it, to ensure robust collection of data. As Brittain (2012) points out the interview is often perceived as being the 'least reliable, as well as the least fair' means of predicting job performance [401]. This is due to such factors as pre-judgment in which decisions about employment are made during rather than after the interview process. This chapter, although focusing on the inter-relational aspects of interview techniques general, has a specific focus on the cognitive interview which is the form of structured interview that produces the most reliable form of interviewee information but is not used for workplace assessment or job prediction. Thus there is a difference between the conversational or structured interview for the purposes of assessing candidacy for employment, and that which is designed to elicit information in response to a particular circumstance or event. This chapter will focus on the second of these.

Obtaining full and accurate information is crucial to an exercise in which information is needed, or can be elicited or when an investigation is carried-out but the eyewitness literature reveals that complete and accurate recall under questioning from witnesses is difficult to achieve. One of the reasons for this is that subjective judgements about the veracity of an interviewee's recall are widely regarded as only being accurate at 'levels of chance' or a little above [402]. Risks of false positive judgments (wrongly concluding that a truthful person is deceptive) and false negative judgements (wrongly concluding a deceptive person is being truthful) meaning that no single method for eliciting accurate information is likely to work all the time [403]. FCT (forced choice testing in which the likelihood of an innocent person missing all target choices is known, or to the choice of alternatives to which the respondent resists) and MCI (modified cognitive interviewing) useful for detecting deception in low-base rate circumstances [404]. As Windle, McFann and Ward (1955) once put it, "[t]he problem of how to most efficiently evoke responses which the interviewee is prone to suppress or repress is probably the basic problem that interviewers have attempted to solve" [405]. Conversely, the effects of 'verbal overshadowing' on witness memory performance is not deemed a concern in research carried out by Brown, Lloyd-Jones, and Robinson, (2008) et al., unless elaborate interview techniques are used. However, 'verbal overshadowing' is troublesome at lower thresholds of information recall veracity because it may introduce misinformation [406].

British police use the PEACE interviewing framework which is a mnemonic for the Interview stages (Planning and preparation; Engage and explain; Accountancy, Closure and Evaluation). PEACE has two interview styles – the conversational management style which is used with resistant interviewees and the cognitive interview which is used with co-operative interviewees [407]. The main tool for law and enforcement agencies is the cognitive interview (CI) which was developed by Fisher, Geiselman and Amador (1989) from requests by Police Officers on improving witness recall. The procedure for the cognitive interview basically comprised four techniques and some strategies for communication improvement [408]. These are based on the reinstatement of the physical and personal contexts of the time of the event.

The context reinstatement involves firstly emotional elements (how was the interviewee feeling at the time?); secondly, perceptual features, an examination of what was perceivable to the senses; thirdly, sequencing the elements of what the interviewee was doing at the time, including rationale for context reinstatement, encoding specificity principle, instructions for extensive search for details which lead to the recall of additional relevant information [409]. These lead to the recounting of events in different orders according to a variety of perspectives. Refinements of the model include the 'cognitive techniques' for activating and probing a witness's mental recollection in image of the various sections of an event, such as suspects face and clothing and objects. Cognitive interview techniques make a distinction between conceptual image codes (an image stored as a concept) and pictorial codes (the mental representation of an image). However, Davis, McMahon and Greenwood (2004) argue that interview technique should not vary according to the imagery ability of the interviewee however some components of the cognitive interview (CI) such as context reinstatement are more impactful [410]. Uninterrupted free recall and some scepticism regarding the accuracy of direct questioning is important, as are the socially facilitative features of the CI (cognitive interviewing) such as empathy, control of answers and rapport building are important [411].

Social communication techniques have also aimed to transfer the control of an interview from the person conducting the interview to the witness. This allows the witness to dictate the pace of the interview and to order their own recall through the use of questions, question timing, and by not interrupting witnesses, so that the recall by the witness is related to their retrieval patterns and not necessarily to the protocol that the interviewer is using. The implication is that the cognitive interview may be (structurally messy) but generates a higher degree of useful data and information recall. The revision of the cognitive interview thus involved the introduction of social techniques which included rapport building, transferring control of the interview to the witness, emphasis on open-ended questions, and use of guided imagery during questioning [412].

Cognitive interview techniques can result in an increase in the correct amount of information but also an increased amount of confabulation. However, Dando, Bull, Ormerod, and Sandham (2015) argue that a gradual disclosure technique can enhance the detection of deception in interviewees in that it asks the interviewee to account for each piece of information prior to recognition of the import of its nature, or under direct challenge, thus limiting the opportunities for the interviewee to piece together an alternative (incorrect) account or to verbally maintain control over verbal performance according to a consciously fabricated script [413].

Some interview techniques such as ". . . repeating pointed, specific questions and applying social pressure, providing negative feedback, and encouraging witnesses to 'work harder' or to give more correct information" do not generate good results because they lead to reporting that is changeable and so not accurate and credible [414]. On the other hand, it is advisable to place the interviewee or witness at the centre of the interview by establishing rapport, reinstating the context of the event or topic, encouraging free recall, transferring control to the interviewee [415].

Repeated questioning may result in negative outcomes. These may include over-confidence, confidence influenced credibility, inconsistency, in response to witnesses changes in answers which may or may not be accurate, and repetitive recall may exacerbate the incorporation of post event information, retrieval induced forgetting, and uncertainty speculation [416]. Consequently, suggestibility in forensic interviews has two components. These are misinformation and shifting the responses from negative feedback under which conditions true responses shift to negative responses. Indeed, repeated questioning is a form of negative feedback [417]. Questions are delineated to the extent that they are answered. There is a division between answerable and unanswerable questions. Unanswerable questions ask about information that wasn't present, except they require rejection or fabrication (209) [418]. As Wsyman et al., (2014) suggest, "CI [cognitive interview] can serve to protect against inconsistency in the face of repeated questioning" if no negative feedback is provided [419]. There is also limited and mixed evidence in interview contexts of the advantages of visual aids, which in some circumstances may allow an increased free association of recall with the tacit knowledge that viewing pictures or objects may produce cultural conditioned responses [420].

There is also some limitation in the research which has been conducted on cognitive interview (CI) in that it has primarily examined its effects under optimal encoding condition (for example, in a single interview after a brief delay) whereas in the real world a witness can be interviewed many days after an event. There have been two main positive effects of this and of repeated testing. The main factor is that it produces reminiscence a recall of previous unrecalled or undiscovered material. It can also lead to a hypermnesia effect, in which the new information exceeds the amount of information which is forgotten [421].

The cognitive component of the cognitive interview (CI) involves the application of four mnemonic strategies for recalling explicit memory processes. Explicit memory involves active recall of a specific episode from experience. No contrast implicit memory can occur without a specific intention to recall. Implicit memory does not rely on previous held knowledge or conscious recollection.

The first of these four mnemonic strategies is, the reinstatement of the context. This is a condition placed on memory recall to provide additional cues for retrieval by reinstating feature which were present at the original time of encoding. This process can be one of internal regeneration and give rise to perceptions that are described from the process of re-imaging – an analogue to remembering a sequence of actions [422]. The second strategy involves the reporting of everything, which is irrespective of subjective confidence. Confidence is a reliable indicator of the accuracy of information but participants in this process are asked to disregard their confidence in, the perceived importance of a piece of information.

The third strategy involves recalling the events in a different order; changing the order of recall should promulgate the use of different retrieval information. The fourth strategy, involves a change in perspective, the recalling of an event from another's perspective which may bring about a change in retrieval description and the making of new information available under intentional learning instructions producing a shift in perceptive which influenced the recall of details of a narrative text, which may operate outside of an individual's awareness [423].

The phenomenological experience of the imaginal perspective may alter under such conditions according to research by Nigro and Neisse (1983) [424]. This may produce an effect in which a person may 'see' him or herself in the memory (from the observer's perspective) or he or she may imagine the event as originally observed (from a so-called 'field perspective"). Observer perspectives may require a person to adopt a frame of reference that he or she didn't experience or to import information from other sources – such imported information can appearance congruent with an event, although information was not contained in the original event. Memon et al., (1997) note that new information may exceed the amount of forgetting of old information. [425]

Factors for consideration under these conditions derive from the interviewee's personal involvement with an event – emotionally content may be remembered differently, phenomenological aspects may change – rehearsal of details produces more vivid images; confidence of memory for some details is greater than others, and control and interference of memory may vary. Such influences may vary at the time of experience (encoding) and at the time of remembering (retrieval) [426]. Furthermore, personally significant events may be incorporated into information about the self, which may influence the retrieval cues available. Personal; involvement in the event may also have negative consequences such as enhanced distress in recalling memories that are painful.

Interviewer's also need also consider the 'uniqueness of the event' – how much more knowledge or familiarity to the interviewer has about what is being recalled? Such knowledge under recall may be influences by the regularity of the event. Experience may influence the ease of retrieval. The 'density of the event' is also according to Gesielman et al., (1985) a factor for consideration [427]. This involves the number of actions contained in the scenario under recall, how fast that conflicting details of information may concur, and the greater the effects that were found with scenarios having many simultaneous actions at once [428]. Showing disagreement may give rise to the kinds of truth-searching behaviour needed for accurate interviewee reporting [429].

The basic construct of recall, are those involving seven questions of – who (which persons); what objects, what verbs, when, where, how and why. Interviewers need to consider the centrality of plot relevance and perceptual and spatial distinctions. But finally a cautionary note is made that errors produced under cognitive interview (CI) will be more resilient than errors produced under less rigorous interview methods. However, research by Smith and Bull (2014) has shown that the value of forensic evidence and the relationship between perceived evidence strength and the disclosure of evidence in police use of interviewing leads to situations in which the significance of evidence elicited may be ambiguated. Hence awareness and interpretation of forensic evidence in the cognitive interview situation is also important [430].

A further non-coercive interview strategy is known as the Scharff-technique, which is intended to elicit intelligence in a friendly manner. It is characterised by four features – a friendly approach; not pressing for information (with the absence of explicit questions); the illusion of 'knowing-it-all,' tacit is that the interviewee is unlikely to know more information than possesses, but might do so if provided with more context. The fourth factor is the confirmation/disconfirmation which involves the telling of a story in which specific features either get affirmed or disconfirmed, hence eliciting information without the interviewee's knowledge of the objectives [431].

DISCLOSURE OF PERSONAL INFORMATION

It is generally understood that disclosure of personal information is central to any inter-personal communication process but what is self-disclosure? According to Chaudoir and Fisher (2010) it is "the sharing of personal information with others through verbal communication" and as such as an intrinsic part of social behaviour [432]. Inter-personal disclosure allows people to develop and express a sense of self and to build intimacy in personal relationships; as such it is linked to individual self-well-being [433]. Disclosure may or may not inform professional contexts because it can alter boundaries of detachment, anonymity, equanimity and abstinence that might be associated with the former in some contexts [434]. However, a process of self-disclosure is integral to any interpersonal communicative exchange that is meaning bearing and in counselling and therapy is a necessary step for improving well-being. This chapter will discuss six facets of disclosure: group disclosure, emotions and disclosure, ethics in disclosure, transparency in disclosure, online disclosure and growth in disclosure.

Disclosure in Groups

The characteristics of disclosure in a dyad or group context may involve the following factors: deconstruction of positions of power within a dyad or group experience, the strengthening of a bond, the normalisation of a patients difficulties, the challenging of negative behaviours, enhancing positive expectations, exemplifying desired behaviours, a propensity for selling, transfer and exchange, positive and negative experiences, sharing of opinions, intersubjective intimacy, and the sharing of demographic data [435]. Self-disclosure can also involve the personalisation of autobiographical fact, the sharing of feelings and insights into past experiences, approval, legitimization and challenges to thoughts and behaviours. It encourages participants to open-up and encourages psychological relief and a sense of well-being [436]. Disclosure between people also implies a form of give and take, a form of 'mutuality' and also that opening-up of self-revealing discussion between people can be a form of imitation. It can also be a form of reinforcement which validates the discloser's sense of reality [437]. For disclosure to be effective it needs to have the property of 'congruence' which is a correspondence between the conscious self and the phenomenological field in which he or she is situated. Incongruence can lead to maladjustment and form of disorder. Consequently, self-congruence is also important, the alignment between self-expression and self-enhancing experiences which lead to autonomy and growth – when symbolic focus is not focused one can't distinguish regressive behaviour from growth [438].

Emotions in Disclosure

Emotional disclosure, the sharing of personal information frequently has a number of positive benefits [439]. These include an increased immune system response, decreased medication and pain, decreased tendency for depression, stress and anger, emotional intensity and reduction of intrusive thoughts, concealment in comparison is thought to produce negative results, however if discourse results in shame or alienation, it should have avoided, despite the fact that in general disclosure is beneficial, depending on the quality of reaction. There is also a high negative correlation between emotional disclosure and suppression of emotions. It is the goal of disclosure to decrease negative emotions [440]. It stands to reason therefore that individuals with impaired emotion may not emotionally disclose to the extent that they benefit from doing so as much as those who do [441].

Attachment to others is related to emotional regulation and disclosure. Brennan, Clark and Shaver (1998) identify adult attachment patterns as occurring along a dimension of attachment avoidance and attachment anxiety [442]. Those who are high in attachment anxiety have close ties to others and fear abandonment; those high in attachment avoidance usually self-reliant and have a difficult time trusting others [443]. Insecurely attached people are high in attachment avoidance or attachment anxiety whereas individuals who score low on these dimensions are securely attached [444]. Distressed emotions may produce high attachment anxiety engage in hyper activating strategies which gain proximity to partners as individuals' response to need, exaggerated subjective negative emotional reactions to minor threats and ruminate over negative emotions [445]. If self-disclosure is primarily the relation to others of the more stressful events that make up life experiences in the attempt of recovery, as Foynes and Freyd (2013) put it, " . . . research indicates that responses to disclosure, rather than the act of disclosure in and of itself, are strong predictors of post-disclosure outcomes" [446]. Thus while there may be increased anxiety associated with disclosure itself and also a corresponding relief, it is the reactions to disclosure (positive or negative) that predict the quality of the outcome.

Ethics and Disclosure

If self-disclosure is the sharing of personal information (which may include personal background, personal history, feelings and reactions) then disclosure may also involve boundary transgression [447]. In some situations, this is healthy and in some situations (depending on the type of disclosure) it may be inappropriate. The positive feelings of increased self-esteem that result from disclosure do so more commonly from people who disclosed for ecosystem or 'other-focused' reasons. One reason that 'first-disclosure' experience is related to current well-being is because it may function to lesson chronic forms of fear of disclosure [448]. Once individuals decide to disclose, the disclosure and accompanying behaviour and responses can affect a wide-range of psychological health indicators including, stress and illness progression or even remediation. However, interpersonal disclosure is usually only beneficial when it elicits supportive responses from confidants [449]. The intentions of disclosure may include social support, strengthening of relationships and professional treatment [450]. The most popular forms of motivation for disclosure are 'eco-system' motivations which are the default or 'other-focused' reasons and affect both the self and the disclosure confidant to yield outcomes which include the strengthening of personal relationships and avoid undesirable outcomes such as stigmatized identities [451]. Consequently, first-time disclosure events may impact peoples' long-term psychological being because they are belief-affecting and the degree of fear of disclosure is informed by prior disclosure experiences [452].

Transparency in Disclosure

Disclosures are intended to regulate the gap between the informed and uniformed can be a burden to those it intends to protect but if advice from a discloser is not trusted the propensity to comply with it may be reduced. There are two factors that may affect this reduction: firstly, if disclosure is provided by an external source; secondly, if the matter of disclosure is not common knowledge; thirdly, if the advisee has an opportunity to change his or her mind; fourthly, if the advisee can make a decision in private about the matter [453]. On the other-hand 'moral licensing' can free advisors from need to reconcile personal interests and professionalism which may produce biased advice in some situations [454]. The disclosure relationship may also be compromised by a conflict of interest which is a conflict between an individual's professional role and his or her personal interests [455]. In this situation an advisee may adjust for the biasing influence. When alerted to conflict of interest recipients may be less satisfied and trustful of the advice they receive because they are less certain that it is given in their best interests [456]. However, a 'pan-handler' effect is observable in which a desire to satisfy an advisor's personal interest is tacit with common knowledge of them, furthermore, situations of reluctant altruism may arise in which people feel pressured to a course action but would have preferred not to take it [457]. The additional pressure to be seen to be cooperative may also override a conflict of interest situation in some face-to-face interactions [458].

Characteristics of Online Disclosure

When individuals present themselves or disclose online they may adopt a variety of self-presentation strategies which are the result of conscious decisions about how to self-present based on people with whom they interact and may consider context, audience and environment as factors of self-presentation [459]. However as Boyd (2008) suggests, the dynamics of social media (online) differentiate it from traditional interaction. These dynamics include the disclosure to invisible audiences, the collapsing of media contexts and the blurring of public and private lives [460]. The users of online social media may combine audiences with disclosures to provide a form of bridging social capital but social media users are also potentially exposing themselves to personal, sensitive and potentially stigmatizing information through profile usages [461]. Personality traits amongst social media users also bear different characteristics. For example, social networks formed by extraverted people tend to be larger although networks formed by introverted people tend to bear more personal information; conscientious people use fewer photographs and neurotic people tend to use more; more information is sometimes provided by people who are open to new experiences; those with high emotional intelligence are able to regulate their content [462].

Growth and Disclosure

For inter-personal relations to inspire growth what is necessary is what Roger's calls 'unconditional positive regard' which can lead to actualisation by removing the environmental threats (such as internal inconsistencies) to congruency which enable a restructuring of the self-concept (Standal 1954; Rogers 1961). Incongruent or defensive behaviours occur in people who have a lack of self-regard as well as regard for others. Self-accepting or congruent behaviours are more readily accepting of behaviours of others. The characteristics of a helping relationship which contains the behavioural ingredients of other-produced growth include [463]:

- Trustworthy communication
- The expression of unambiguous separate selves
- Positive characteristics of warmth and caring
- The maintenance of separate identities
- Empathy
- Acceptance of other's experiences
- Sensitive responses
- Threat-free evaluation
- The permission of change

Consequently a growth-cycle necessitates a recognition that: firstly, the world (or even a person's world-view) has changed; secondly, that a present experience has shattered a 'world (-view creating) structure'; and thirdly, that the retotalization of the world structure (view) encompasses the idea of a changed reality [464]. Thus, on balance disclosure is a positive attribute but when examined in depth it has both positive and negative features.

This chapter has provided definitions of disclosure and briefly examined six characteristic facets of it in order to highlight its on-going consideration as a mode of interpersonal communication in the counselling context. Interpersonal communication in counselling cannot happen in dyads, for individuals or groups without some form of disclosure taking place. Disclosure is the ‘bread and butter’ of inter-personal communication in any counselling or therapy exchange, it therefore deserves continued reconsideration as the initial principle mechanism of counsellor-client conversation.

Chapter 11

THE PERFECT BLUSH

Human emotions regulate our thoughts, feelings and perceptions, reflections and interactions with others and the environment. Human emotions are forms of physiological arousal that have a neuro-chemical basis and a qualitative self-reflexive characteristic expression which governs bodily expression and somatic sensation (See Figure 7). Human emotions are on a continuum or spectrum that has positive and negative qualities. We can be joyful or sad, ebullient or soulful, and the ranges of emotions have an almost infinite variety of intensities and accompanying thoughts and feelings, within the imprimatur of the basic emotional 'type' – anger, sadness, joy, anxiety, and so forth. Emotions are like the 'physiological' 'skins' we inhabit according to how we perceive our relationship with the environment and with others.

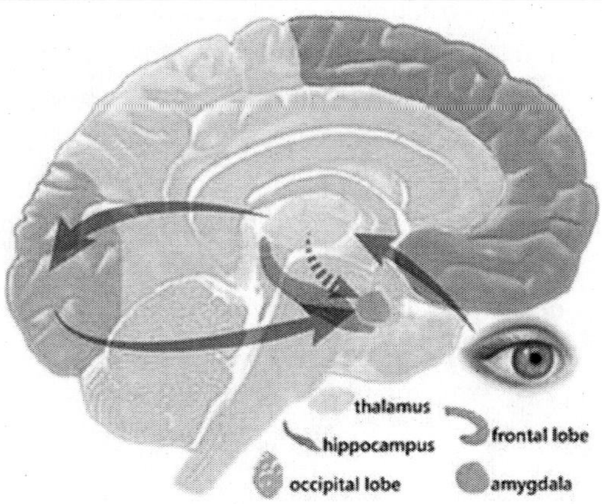

Figure 7. Centers of emotional stimulation and regulation in the human brain.

The centre of emotion is the amygdala which is one of the earliest brain organs to develop. Branches of the humanoid family have felt some form of emotion – in the most basic form either anxiety or pleasure, probably for as long as they have had 'second order' brain organs. Emotions, based in the amygdala precede but are nevertheless intimately tied to our complex physiological arousals and neocortical stimulations. Goleman (1996) [465] coined the term 'Amygdala hijack' to describe emotional responses that are acute and overwhelming as a result of emotional threat. Neurochemical transmissions from the thalamus go simultaneously to the amygdala and the neocortex, these are cross referenced with a store of experiences in the hippocampus, then a stimulus is sent from the amygdala to the HPA (hypothalamic-pituitary-adrenal) axis and the ancient existential response raises the adrenaline and the rational brain is 'hijacked' for immediate response. Arguably, because of the 'safer' boundaries of modern developed societies and the relative peacefulness of everyday life that it affords, the basic flight or fight response of people, the stimulus that makes people anxious, has become less necessary. But how long will societal conditions that have made it so last, long enough for there to be some evolutionary modification or selection away from the intensity of this brain function? Might we slowly be selected for other functions, such as empathy and an ability to think of others? This chapter will focus on two emotions that are central to the everyday perception and modification of the working world, embarrassment and confidence. The former an emotion of excessive self-monitoring and the latter an emotion of the relaxing of self-monitoring.

Embarrassment is a self-conscious emotion, an emotion that people may experience when they have violated a social convention or disrupted on-going social interactions. The anticipated anxiety and subjective qualities of embarrassment lead people to avoid it. Anticipated anxiety as a behavioural state can impair everyday social interactions [466]. The antecedents, experience, and display of embarrassment are distinct from guilt, shame and amusement thus they qualify as separate emotions [467]. A further definition of embarrassment is that it results from being the centre of attention of something foolish, social misfires, *faux pas* or 'sticky situations' that challenge conventional roles [468].

EMBARRASSMENT AND EMPATHY

Empathy is a vicarious affective response which strongly matches another person's emotional state – reactions range from sympathy and compassion to discrete emotions that match a person's expressed feelings [469]. Individuals may consciously engage in perspective-taking or imagining themselves in another's place and experience an empathic emotional response [470].

Function of Embarrassment

Although embarrassment can be unwelcome and people often seek to avoid it, it is nevertheless claimed to serve various social functions. Observers of embarrassment in social situations have led behavioural psychologists to believe that it can "serve to mend social relationships following an accidental violation of social expectations" [471]. If people violate social norms they may undermine rules of demeanour that are the substance of smooth-flowing interactions. A display of embarrassment is not a dysfunctional, irrational outburst but is a patterned expression that restores social relations gone awry, and signifies an individual's concern over norms that provide the foundation for social relations and future commitment to norms [472]. Darby, Harris and Harris (2006) state that there are three functions for embarrassment [473]. The first is appeasing the offended, the second is deterring repeated transgressions and the third is motivation for the embarrassed to repair the social rift caused by the blunder. The overriding function of embarrassment is to gather emotional information from an audience, as opposed to a socially anxious person's reaction of withdrawing from emotionally laden social situations.

Body Language of Embarrassment

On an index of behaviours, individuals who experience embarrassment often display gaze aversion (shifting positions, speech disturbances), smile controls and inhibition, a tendency towards downwards and sideways head movements, and laughter and nervous face-touching [474]. Participants involved in experiments which elicit embarrassed behaviours from individuals subjectively report of feeling funny, awkward, foolish, nervous, surprised, and self-conscious with an inclination to hide or withdraw [475]. However, in comparison to shame, embarrassment is brief, light-hearted and mild – shame is characterised by a feeling of being a bad, immoral person, with long-lasting anger and disgust with self, a feeling of isolation and a propensity to apologise more profusely than might be the norm. Embarrassment is greeted with surprise but shame with emotions of anger, contempt and disgust. A blush, for example is thought to be a sign of moral character [476].

Experience of Embarrassment

Because of the negative subjective qualia of self-exposure which produces embarrassment, people often try to avoid experiencing it. It leads to anticipated anxiety which can impair everyday social interactions and lead to disorders. Whilst the experience of embarrassment is painful, it can serve social functions such as mending social relationships after violation of expectations [477]. Feinberg et al., (2012) reason that a transgressor signals a commitment to the social order and re-orientates towards it in a pro-social fashion. It demonstrates a concern over the potentially adverse consequences that a violation could have on others, and an individual has a tendency to want to be seen as an upstanding member of s social group [478]. Embarrassment is similar to the experience of social anxiety and shares a propensity to avoid negative feedback (such as angry eyes). The experience of embarrassment possibly provides a motivation to look for social feedback in order to reconcile with an audience [479]. Though is seen as inconsequential, it can have a strong effect on social behaviour. Embarrassment probably is also a distinct emotion, the antecedents, experience and display of embarrassment and its autonomic physiology are separate from other emotions such as shame or guilt Embarrassment is also a public emotion and involves some impersonal interactions and implicates an outer public self rather than an inner self [480].

Mimicry and Embarrassment

Mimicry is the outward display of empathy. Empathy is an affective and frequently vicarious reaction to another's emotional state or situation; it may range from sympathy and compassion to discrete emotions that match feelings or circumstances [481]. Mimicry of another's nonverbal emotional expressions can influence one's congruent effect [482]. Hawk et al., (2011) also state that mimicry can be seen as a part of perspective-taking or imagining themselves in another's place. Empathetic embarrassment can result from conscious cognitive processes in the absence of a target emotional expression; emotional displays which elicit perspective-taking can serve as a mediating route to empathetic embarrassment [483]. However, there is also a possibility that mimicry is more of an automatic or unconscious process.

Causes of Embarrassment

Displays of embarrassment are frequently not irrational but they occur when individuals become self-consciously aware that they have violated certain self-conscious norms rules of deference that produce smooth-flowing social interactions [484].

Associated with violations of social conventions, heightened concerns of social exposure and violation differ from antecedents and appraisals of other emotions such as shame. The 'loss of self-esteem' account of embarrassment posits that people feel embarrassed when they have failed to act in accordance with social standards which leads to a transient state of depressed self-esteem and self-disappointment [485].

Models of Embarrassment

There are at least six models of embarrassment. The loss of self-esteem model holds that embarrassment results from a decline in self-respect or dignity, which follows from negative evaluations of others, whereas the social evaluation model argues that an undesired evaluation will result in embarrassment [486]. The loss of self-esteem model thus argues that embarrassment results from a change in perception following an evaluation of the self in relation to others. By comparison the personal standards model, argues that embarrassment results from a discrepancy between one's behaviour and one's ideals. The dramaturgic model of embarrassment emphasises the evaluations between self and others and holds that embarrassment results from the disruption of a social interaction due to an inability to act in a given social role. Similarly, the transgression of others' expectations model places emphasis on the evaluation of others; a person praised by others and not embarrassed by a supportive evaluation but embarrassed because they are not sure how people expect them to respond to praise [487]. The centre of attention model posits that being the centre of attention can simply be embarrassing, even when there is no loss of self-esteem of failure to act. The unwanted exposure model by comparison posits that embarrassment follows from exposure to something a person prefers to conceal. The relationship of professionalism to the experience of embarrassment is evidenced by how it produces pro-social emotions; it facilitates the smooth functioning of social activities - it is a regulator of social activities and it provides a check on the perceptions and the veracity of individual and social information.

CONFIDENCE

Feeling supported and ready to take on the events of the day. The blind courage that enables you to carry on and find a way even when you are struggling. The ability to lift a mood in a room full of anxious people. The ability to cross a swinging bridge without fear of falling. The resilience to comfort an ailing child. The ability to select the right tool for the job. All these are characteristics of a person who has confidence, a feeling of internal resilience, or a feeling of personal ease, and of being up to the task and more. Some people seem to have confidence in a fuller measure than other people. We look and why and when this may or may not be a useful thing.

Fridja's (1988) first law of emotions states that, "events that satisfy the individual's goals, or promise to do so, yield positive emotions; events that harm or threaten the individual's concerns lead to negative emotions" [488]. Satisfactory emotions lead to confidence. A confident self-view applies to the degree of certainty that one has in one's concept of the self [489]. Why this is important is because many of life's challenges require confidence. Confidence is measured in degrees – simple everyday living requires a degree of confidence. Confident self-views are formed from coherent evidence of the self - gained from knowledge of the environment and past experience, they require stable evidence and are influenced by fluctuations in the immediate situation. If an individual's self-view is threatened it may result in a temporary loss of confidence and in behaviours that aim to restore confidence in a self-view. Some symbolic behaviour may be involved in recovering a self-view of moral worth – a behaviour that enacts a rite of passage into a more moral self-view – termed the 'Macbeth effect' for its indirectness [490]. Subtle features of situations or actions may trigger metacognitions, signalling that something is wrong, which results in reduced confidence in thoughts related to a consistent world view. However, these may be bolstered by both direct strategies (which aim to boost confidence in the threatened self-view) and indirect strategies (affirming a similar self-value) [491]. Thus self-view can not only be manipulated by the content of one's thoughts (related to being moral or immoral) but also can be influenced by subtle factors [492]. Individuals desire to hold self-confidant views but confidence can be aversive which produces negative psychological outcomes which include low self-esteem, unhappiness and anxiety. If confidence in a self-view is cast in doubt, a person may recover from negative psychological outcomes and restore a level of self-confidence [493]. Many different strategies to cope with low confidence may be derived, however a common strategy may be to resolve an uncertainty by obtaining objective information about the self. An indirect solution is to cope with a threat to the self through re-affirming an important self-value. Self-values that have been shaken along one dimension could be restored in an overall self-view by affirming an unrelated self-value. Rationalisation of an event or the minimisation of its value can also reduce threats to the self [494].

People process information more carefully in states of low compared to high confidence [495]. Under moods of low confidence, people focus on concrete construal's whereas under high confidence people focus on abstract construal's [496]. When in states of low confidence people view messages framed in a concrete manner as *somewhat* relevant; in contrast people in a state of high confidence view abstract messages as *more* relevant and so engage in the processing of abstract messages more frequently [497]. High confidence is defined as feeling a state of certainly or surety about one's environment, ideas or actions, whereas low confidence, is a feeling of being unsure and uncertain about the environment [498]. Weary and Jacobson (1997) revealed that when people are uncertain about the causes of outcomes they engage in greater information processing than those who are surer or confidant of outcomes. Similarly, individuals in a state of low power and low confidence are influenced by whether the merits of an argument are strong or weak unlike their more confidant peers [499]. Consequently, it can be inferred that people in states of low confidence believe they have insufficient knowledge for a given situation hence they may compensate by seeking out and processing information. However, high confidence leads people to infer that that knowledge is sufficient and the information processing corresponds to the appropriate level of abstraction as the message [500]. Top-down information processing occurs when situations are perceived to be benign and guided by high level abstract information. In such cognitive environments ". . . confidence should lead to the construction of schematic, decontextualized representations of objects and describing events in terms of subordinate, essential and goal-related features" [501]. In situations when people experience low confidence the focus is on the data at hand and thinking is towards low-level details and contextualised information. As a consequence, these give rise to contextualised representations of objects that include "subordinate, incidental, and means-related features of events" [502]. State of high confidence reduces depth of information processing [503].

When considering the role that confidence makes in judgment and decision making, no issue is more prevalent and problematic than over-confidence. A couple of concepts may apply. These include overestimation – a belief that you are better than you are, over placement – thinking that you are better when you are not and over precision – being overly sure of the truth [504]. Over-confidence leads people to make mistakes. When tasks are easy, people tend to underestimate the competition. When information is lacking, people are likely to be overconfident as they lack the crucial feedback for calibrating their judgment (for example, when we are unfamiliar with a competitor we may over estimate their ability with easy tasks and under estimate on difficult tasks [505]. Although confidence tends to improve social status, people who are perceived as competent may not be the most competent, similarly, overconfident people tend to speak more. However, confidence can backfire when it is exposed as overconfidence. Confidence may inspire emotions that stimulate learning – 'achievement-based emotions", whereas debilitating emotions may interfere with learning and performance. With respect to the development of confidence, several factors have been identified: "social support, coaches' leadership and physical self-presentation – mastery of or the sense of improving one's skill" [506]. Furthermore, confidence was more likely than natural talent to be cited as an attribute for success; indeed, amateur athletes view confidence as important and more valuable than physical strength or accessories [507].

Moreover, confidence is related to decision making. The confident decisions made and the time taken to make them are fundamental components of decision making. Concepts of confidence are used in both perceptual and general judgments and claims about memory. According to Ratcliff et al., (2013) confidence is twice as likely as natural talent to be an attribute for success [508]. Subjective confidence is used in perceptual judgments, however it remains questionable as to what the bases for perceptual judgments are and the reasons for accuracy. For example, confidence in perceptual judgements is qualitatively different from confidence in memory tasks [509].

In the area of judgments involved in decision making, an area of focus is the calibration of assessed probabilities. The overconfidence bias is observed when people assess the likelihood of answers to given questions as being correct. Underlying questions are whether confidence is diagnostic of accuracy; how are metacognitive judgments formed; what processes underlies the monitoring of knowledge; what is the accuracy of metacognitive judgments and the correspondence between subjective and objective indices of knowing; what are the processes underlying accuracy and inaccuracy of metacognitive judgements; what are the effects of monitoring on control, assuming metacognitive judgments exert measurable effects on cognition and behaviour, and how do effects influence performance? [510].

People express confidence judgments that exceed the accuracy of judgments [511]. However, in the expression of confidence in judgments involving perceptual tasks there is good evidence of calibration or the tendency towards under confidence. Perceptual judgments are more accurate than what peoples feel. This tends to be correct even when a participant's task may not be vulnerable to such overconfidence. On the whole, people exhibit an under confidence for easy judgments and a progressive overconfidence as difficulty increases [512]. Confidence decreases with latency (or time passed) in making choices [513].

An overconfidence bias is believed to derive from the failure of researchers to sample items representative of a natural environment – when a representative set of items is used, confidence should be well calibrated [514]. Knowledge tasks concerning calibration involve an absolute correspondence between confidence and accuracy. The second feature is resolution or relative confidence. Resolution refers to within-person confidence and accuracy which reflects the ability of participants to discriminate between correct and incorrect answers.

Self-constancy and choice require represent mnemonic cues derived from experience gained in the decision making process and the amount of deliberation and conflict involved. Autonoetic consciousness or the awareness of personal time (including the past and the future) is characterised by retrieval from episodic memory (i.e., personally experienced events), while noetic consciousness, or learned knowledge that is accompanied by personal awareness, is retrieved from semantic memory. Or there may be separate retrieval processes of differing strength levels along an undifferentiated dimension which relate to different underlying memory systems [515].

Witness high confidence influences jurors even in face of contradictions. Indeed, post-identification feedback, rather than actual quality of memory for the perpetrator is necessary so that jurors can rely on distorted evidence [516]. The attribution process is important in forming legal judgments and a phenomenon called confidence inflation, in which jurors may not automatically discredit a witness testimony can affect outcomes. In this context it is necessary for jurors to think why a witness may become more confident. For example, lawyers may manipulate a juror's perception of confidence inflation to generate the lawyer's own explanation, or jurors may generate their own explanation. In this situation simply introducing eye-witness's confidence to a statement may be insufficient to combat confidence.

When people are given corrective feedback, errors that are committed with high confidence are easier to correct than low confidence errors. Responses held strongly are more easily overwritten, and deeply held beliefs may be more easily corrected – responses in high confidence are those in which a person believes most strongly, or strongest in memory [517]. There are two explanations for this. Firstly, the 'attentional explanation' whereby when people are wrong with high confidence, they are surprised and embarrassed and compensate to correct the item. Secondly, the familiarity account which involves the idea that there are systematic differences in characteristics of general information, related to high confidence errors or individually based differences in the participant's own familiarity with domains as compared with low confidence errors [518].

People claim after receiving the corrective feedback following high confidence errors that people 'knew it all along.' Those items that evoke high-confidence errors may be inside what we have called person's region of proximal learning [519]. Furthermore, the commission of errors does not harm eventual memory performance. Subjective confidence in errors plays a role in which errors are most likely to be amended – those held with high confidence are hyper corrected after corrective feedback. Thus it may be that experiential intensity rather than value is rewarded.

Realistic confidence in judgements about retrieved memories is important in medical and legal contexts. The realism of confidence judgments pertains to how well a person's confidence for a memory report matches the veracity of the report (confidence realism is also called confidence accuracy) [520]. The concept of confidence realism includes two factors – calibration – which is the relationship between the level of confidence judgments and the probability of the answer being correct; and discrimination, the extent to which the respondent can discriminate between correct and incorrect answers by means of their confidence judgments [521]. People show overconfidence for general knowledge questions. Although people show a lack of realism in their confidence judgments in many contexts they can regulate correctness of memory performance if they can choose which items to report. If cognitive tasks are along a continuum from effortless to effortful – termed processing fluency – the subjective experience of ease of processing information serves as cue for people in different judgment tasks; fluency is associated with truth and disfluency with untruth [522]. But people can be both truthful and wrong. Confidant people feel good to be with, they make life fun and easier but sometimes confidence can be an excuse to not think things through properly. What seems right because it is said with confidence may not be in hindsight. A doubtful mind might exude less confidence but greater accuracy than a mind blind to anything but its own conviction.

Chapter 12

COUNTING SHEEP

Why do people dream and why do people generally recognise only parts of their dreams? Does it matter if I remember my dreams or not? What does dreaming (or absence of dreaming) tell me about my conscious life? Why can I never store my dream-content in long term memory? That dreaming is intrinsically associated with memory has been an opinion asserted by many researchers including Nielsen and Stenstrom (2005) [523], Fosse, Fosse, Hobson, and Stickgold (2003) [524], and Lee (2010) [525]. However, if dreaming is consciously recollected it must be that memory is also active at the time of dreaming, and if this is so, then the use of memory from dreaming must be associated with consciousness in the waking state. If a concept of consciousness is conceived as following from a layering of human perception, cognition, and physiological experience, then the brain may be understood as having the potential to produce the non-linear complexity that makes the experience of consciousness possible. The purpose of this chapter inferences about the relationship between dreaming states, waking states, and memory. Dreaming seems ephemeral yet may have a survival function. Does our sleep and dreaming differ from that of our ancestors, is our conscious world in fact the dream of our ancestors from which we are all awaking?

People are not simply conscious (awake) or not conscious (unconscious or asleep) (although they may be) but rather there are differing qualities and degrees of consciousness -- such as dreaming -- implicit in the cognitive frameworks with which people negotiate the experience of their worlds. So-called "lucid" dreamers for example are able to think clearly, to act or reflect whilst experiencing dreaming as LaBerge (1990) has noted [526]. Furthermore, as Hanfling (1984) states, "If a tribe of people to whom dreaming is unknown were to learn our language, they would also have to learn the distinction between dreams and the reporting of dreams" [527]. Conscious states and the 'handle' we are given on them by ordering our thoughts through language are the symbols of our conscious processes and sensory perceptions, they can describe and point at an experience, even lead the mind towards that experience, and yet the conscious experience itself is beyond the littoral language meaning. Although dreams are likely to be the product of schema assimilation (and remembered dreams as if of recollected events experienced in consciousness) during sleep, the purpose of human dreams remains as elusive as the dream episodes themselves. Recognition of dreaming states involves the semi-conscious recollection of memory trace during the sleep consolidation-based stabilisation phase, as such dreaming might be considered a by-product of schema assimilation. Sleep is thus necessary for the consolidation-based enhancement of motor sequence learning [528].

Early modern research from Jenkins and Dallenbach (1924) indicated that the strength of a memory representation or "trace" could be more preserved by periods of sleep compared with time awake [529]. This is because sensory processing during sleep is diminished compared to in an awakened state, whereas activity of memory processes may not be so, so more consolidation of memory can take place. Moruzzi and Magoun argued in 1949 that the function of sleep was to reinstate the activity of synapses, or to regulate brain processing [530]. If so then memory of dreams involves both anterograde (future orientated) and retrograde (past-orientated) memory as a remembered dream sequence involves recollection in the present of a future-orientated past physiological experience.

There are two main states of activity during sleep which are both natural cycles – REM sleep (rapid eye-movement sleep) and NREM sleep (non-rapid eye movement sleep). Typically, NREM sleep is followed by a short period of REM sleep and dreaming occurs during the REM stage. Learning through habitual actions during consciousness are REM independent but as Greenberg and Pearlman (1974) put it, ". . . activities involving assimilation of unusual information require REM sleep for optimal consolidation" (p. 516) [531]. REM sleep is also that which is inductive to memory of dreams upon awakening. As Siegeal (2005) observes, although dreams contain emotions and events that don't correspond to previous days, dreams are not recalled unless they are immediately rehearsed in post-dream waking [532]. A person who awakes after the REM sleep cycle is more likely to remember their dreams than one who wakes after a NREM stage of sleep.

A definition from the second wave of sleep-research, provided by Wamsley and Stickgold (2010), holds that dreaming, "[d]uring sleep, when attention to sensory input is at a minimum, the mind continues to process information, using memory fragments to create images, thoughts, and narratives that we commonly call 'dreaming'" [533]. As such dreams may be "spandrels of sleep" [534]. The spandrel while an epiphenomenal term also instantiates a complex neurological interaction. As Hahn, McFarland, Berberich, Sakmann, and Mehta (2012) suggested, the entorhinal cortex layer III is a mediator for memory consolidation during slow-wave sleep and it "inputs to the hippocampus in temporal association memory" [535]. Furthermore Nieuwenhuys et al., (2008) showed that the parahippocampal cortex, which receives diverse sensory-specific and multimodal cortical information (a highly complex area of synaptic functioning) is also useful for REM sleep (see also Markowitsch and Staniloiu, 2013) [536]. Such alteration of brain-wave functioning in various areas of the brain during sleep is consistent with fluctuations in the quanta of memory consolidation. If sleep reverses the deterioration in performance brought about from prolonged wakefulness, then REM sleep is modulated by the circadian rhythm [537]. Hobson (1975, 2005) suggests that neurons which mediate non-REM and REM sleep are motor pattern generators coterminous with norepinephrine serotonin which is non-gating, allowing for conscious brain activity [538]. Sleep, dreaming and consciousness are thus states on a continuum of brainwave quantum functioning.

Why do people only remember dream episodes versus full narrative experiences? As Lee (2010) suggests, "[l]ack of dream recall [as full narrative memories] suggests the modern emphasis on the significance of waking realities at the expense of oneiric experiences" [539]. It is not simply that natural selection has pre-conditioned us to dismiss dreams on waking but that the cognitive states of dreaming are suppressed on waking experience; as Lee (2010) also states "dreaming constitutes an unbroken chain of memory to the organisation of everyday life" [540]. It is rare that memory can provide unproblematic access to detailed aspects of any particular dream. Dream recall is effected during rapid eye movement (REM) sleep, and the ability to recall dreams is most active during the REM phase of sleep. But dream recall and memory in both traditional and modern cultures are regarded as continuous with conscious experience in the waking world [541]. The memory of dreams in a waking state, Freud termed day "residue," -- acknowledging the fact that memories of previous dreaming experience can effect and have a delayed reaction on conscious cognitions. However, Freud's symbolic language of dreams was eclipsed by the activation-synthesis hypothesis of Hobson, McCarley and Wyzinski (1975) which was a neuroscientific account of dreaming that posits that dreams originate from neural signals in the brainstem generated during rapid eye movement (REM) sleep [542]. For Hobson et al. (2005), dreaming occurs when a sleeping brain attempts to process that chaotic input into its "higher-level cortical circuitry" but it can also occur in non-rapid eye movement (NREM) sleep [543].

As Nielsen and Stenstrom (2005) observe, remembered dreams only very rarely portray complete episodes – this occurs in only one point four per cent of reports, whereas in another study, up to sixty-five per cent of incomplete episodes of dream elements were linked to waking events [544]. However, in negative dream states such as nightmares, dream imagery is episodic. As Lee (2010) states, a traumatic event may be replayed as a group of "isolated spatio-temporal, perceptual and emotional details which may or may not preserve autonoetic" (self-in-time) awareness [545]. However, it is clear that (like some forms of imagination) dreams simulate reality in so much as they take place in spatially coherent environments, in which a self-interacts perceptually. There is orientating sensory information, and a sense of self which engages in emotional and intellectual exchanges. Consequently, the spatial and temporal valences that signify dreaming are coterminous with the spatio-temporal binding that characterises consciousness [546].

There is a consensus forming amongst some theorists that altered hippocampus function during sleep accounts for the episodic dream memory and that hippocampal changes (or activity of the hippocampus, entorhinal cortex and other parahippocampal regions) contribute to the characteristics of dream content [547]. Thus when sleeping but in a dreaming state, the perception of a dream as occurring in the here and now may or may not correspond to the spatio-temporal location of the dream. This in itself might occur in parallel with other functions such as narrative organisation, and could occur on the threshold of consciousness and thus be self-observable by both dreamer and awakening sleeper [548]. Such experience at the "threshold of consciousness" could be understood in quantum terms. The hippocampus is responsible for both temporal and spatial patters (specifically the presubiculum regions). However, the emotional sources of dreaming are regulated by the amygdala which controls encoding and the retrieval of emotional memories and physical expression [549]. Amygdala activity raised during REM sleep rather than during wakefulness maintains a reciprocal dependence with hippocampus in storage of memories – amygdala gates sensory information through the entorhinal cortex [550]. The cognitive-level replay that is characteristic of dreams corresponds to reduced levels of acetylcholine in NREM sleep or quiet wakefulness and is believed to result in the consolidation phase of episodic memory, producing information flow from the hippocampus to the entorhinal cortex [551].

Although it is apparent then that there is a link between memory and dreaming, there is as yet no causal linkage. However, when an integrated episodic memory, is experienced during waking, information may flow from the hippocampus to the cortex -- such hippocampal outflow is blocked during REM, with neural information flowing from the cortex to hippocampus [552]. Dreaming during REM sleep results in the suppression of brain chemistry which registers stress and difficult emotional experiences may be processed which results in the ability to recall difficult memories without trauma [553]. Thus dreams may help to emotionally regulate traumatic experiences. As Wamsley and Stickgold (2010) state, dreaming is:

> . . . the product of a mind that is constantly encoding and processing information about the world. When sensory input is at a minimum, newly formed memory traces are stabilized during offline states of quiet wakefulness and sleep, though the repeated reactivation of experience-related activity patterns. During sleep, this reactivation of memory traces contributes to the imagery, thought, and narrative of dreaming [554].

Both REM and NREM sleep are involved in the consolidation of different forms of memory. Consequently dreaming is "influenced by the retrieval of recent memories in the sleeping brain" [555]. Thus dreaming states may be "quantum, trace, or ghost, visions" of the disorganized contents of recent recollected memories, undergoing re-organization in the brain. However, while the relationship between memory and conditioning is not clear, what is clear is that the hippocampus plays a role in the acquisition and retrieval of fear conditioning in animals but conditioning itself may be distinct from human cognitive memory as it produces a far stronger physiological and behavioral response. As Wamsley and Antrobus (2009) assert, "hippocampus-dependent learning can be reactivated and expressed during human sleep, effecting the emotional quality of experience" [556]. Thus, for Walker (2005), given current understanding of the neurophysiological substrates of memory forming and consolidation, "we are able to move away from the question of whether sleep is the key factor responsible for memory formation, and instead, begin disentangling certain confusions around the argument of exactly what type of sleep is or is not required with regard to discrete stages of memory development" [557]. It stands to reason that the plethora of sensory stimuli received in the course of wakefulness will require periods of relative cessation in order to be processed into schema that are memorable and those that are commonplace and can be discounted from conscious recollection.

Tononi and Cirelli (2005), argue that the benefits of sleep, including performance enhancement, are associated with synaptic downscaling as a continuation of the homeostatic regulation of slow wave activity [558]. Such wave activity may also be analogous to a reduction in quantum fluctuation in the "many minds" theory. If it were possible that dreams are thus recalled fragments of memories at the cusp of changes in quantum states in consciousness during REM sleeping states this also may lend credence to the idea that problem-solving as a part of schema processing can be accomplished while asleep. Maestes and Rude (2012) also argue that "over general memory" results in a relative inability to imagine the future—functional magnetic resonance imaging (fMRI) evidence reveals an overlap in neural pathways used in recalling past events: "[F]rom the evolutionary perspective, a primary function of the episodic memory system may be to store and retrieve specific autobiographical memories in order to predict or imagine the future" [559]. Imagining the future is precisely what writers do when they choose words from a lexicon of possible choices to form sentences. This leads to two other hypotheses about memory systems that the writer may need to overcome: functional avoidance and rumination. According to the functional avoidance hypothesis, individuals may truncate their memory to avoid the negative effect of recalling specific memories. In the rumination hypothesis, during the initial cue elaboration phase of memory retrieval those using their working memories may over-process or over-ruminate. This interferes with their ability to recall events or match schemas in the present [560]. Given these hypotheses, an attitude of disinterested openness might favor the writer in their craft. Memory is fallible. For Schacter (2001), memory has seven main problems: transience, absent-mindedness, persistence, bias, misattribution, suggestibility, and blocking [561]. Loughran (2010) makes the point that our lives lived in the present are informed by our concepts of the future and of past experiences [562]. Are enjoyment of the transitoriness of life's brief moments is made more complex by our awareness of possible pasts and futures – not so for all of the animal kingdom.

Chapter 13

AUTOMATIC THOUGHTS

This chapter explores the phenomenon of unconscious or non-conscious communication, semantic information exchange and problems of subliminal communication. That communication occurs beyond spoken words and independently of conscious intentionality and creation is a phenomenon which has concerned communicators, psychologists and behavioural analysts, for decades. This paper aims to explain contemporary understanding of its occurrence and to point to ways in which aspects of unconscious or non-conscious communication can influence behaviour and memory. In so doing it attempts to bring non-conscious communication more into conscious recognition. It will also discuss the concept of the semantic internet or Web 3.0 in the context of interactive unconscious or non-conscious communication. Beginning by offering definitions of unconscious or non-conscious communication, describing psychodynamic communication processes and in considering semantic exchange within systems and human psychology; and finally subliminal communication; the chapter proceeds to critique current knowledge. It demonstrates that one of the issues of artificial intelligence stems from the layer of communicative processes that occur in human behaviour from a secondary field of sensory signals. The chapter underscores the fact that unconscious communication deserves greater study and attention in the workplace in human communication in general.

INTRODUCTION AND DEFINITIONS

According to Pigni (2006) symbolic communication is "unconscious communication" that occurs beyond words, and also the use of silence in conversation, usually on the part of the listener, allowing the speaker to open up, and feel respected [563]. Symbolic communication may or may not be direct, as for example in the use of hand gestures to indicate movement. It may also be pieces of information that we are not aware of, peripheral traffic signals, and peripheral attentiveness from people in a crowd or in certain proximity to oneself. Peripheral subtle communication may not register in consciousness, for example messages flashed on a screen for four-milliseconds may be perceived without conscious awareness. If and when communication does proceed from the unconscious, then people may use more than one mode of communication, saying one thing in verbal communication and communicating another through their voice tone, or gestures and behaviour [564]. So from a 'listener and an observer's' point of view important information and detail can be found in attention to silence and to subtle body language as well as spoken words in conversation. Indeed, in setting Sands (2010) observes that, "... moments of dissociative unconscious communication can give us dramatic glimpses into the relational dissociative unconscious" [565]. Dreams are also a source of meaning involving dual-modes of cognition, imagistic and sensory-dominated and well as linguistically encoded. But dreams only have at best symbolic relationship with reality as residual products of memory trace for ordering patterns for waking experiences. Metaphor activates more brain centres than any other type of communication (Modell, 1997) and metaphor is central to many communicative practices [566].

UNCONSCIOUS COMMUNICATION AND HUMAN PSYCHOLOGY

Jacobs (2001) points out that nothing is lost in the brain, that all memories including earliest ones are registered in neuronal memory banks [567]. So too with communication, some is so subtle as to be barely perceived, such as barely perceptible shifts in demeanour, facial expression, bodily movements, affect or tone of voice or fleeting changes in facial expression, variations in tone, inflection, vocal quality and minor alternations of frame [568]. People are unconsciously exposed to happy and angry faces, people nevertheless react with distinct facial reactions that correspond to happy or angry faces, positive and negative reactions are unconsciously evoked, particularly important aspects of emotional face-to-face communication occur on an unconscious level [569]. Deecke (2012) argues that consciousness is not an epiphenomenon but a brain function or state [570]. Between the information flow in our senses and that in consciousness, selection – sophisticated selection (intelligent filtering, concentration on important matters) unconsciously organised, enormous compression of information, data compression of magnitude ten to the power four.

In the information processing theory of the brain, channel capacity measured in bits, through receptors to nerves of at least ten, to the power of five, bits are flowing [571]. As Zajonc (1980) points out there is a hierarchical organisation of nervous system, which builds a new system with inhibitory properties of old one, evolution of neo-cortex important [572]. According to Zajon (1980), the initial response to stimuli is generated without conscious cognitive awareness, and elicited without involvement of conscious processes [573]. Knoblauch (2005) argues that sub-symbolic processing accommodates infinitely fine variation; this processing is not represented by standard metric systems of computational rules. Examples include changes in emotional states of others based on perception of subtle shifts facial expression or posture, changes in own states based on somatic or kinaesthetic experience [574]. More specifically these might include changes to face and posture, subtle shifts in vocal tone, rhythm -- multiple physical modalities – pelvis, across, heads, gazes, facial expression, voice intonations, and gestures. Mirror neurons were originally discovered in the inferior frontal gyrus (F5 area) and inferior parietal lobule of the macaque monkey (Di Pellegroinio et al., 1992). As Lacoste-Badie and Droules (2014) state: "[m]irror neurons are a particular class of visuomotor neurons in the brain that show activity when an individual performs an action and when he observes another individual performing the same action" [575]. Thus they are empathetic or at least para-sympathetic with environmental stimuli of a certain kind. Although the metaphor of a 'mirror' in the mind might seem simplistic (and might be something that might be better associated with narcissitic personalities). The qualities and degrees of representation that cause mirror neurons to be excited have yet to be mapped but such a class of neurons does afford the possibility at least for a form of human meta-mind interactivity. If so-called mirror neurons did exist and did inspire a form of cognition it might be a window on both a higher intentional order and also inter-subjectivity. Indeed it offers the possibility that in seeking that which is alike, there is great power in the communication of emotion. As Deecke (2012) argues, the Platonic-Christian idea is that the activity of the mind precedes that of the brain, but since the mind is envisaged as immaterial – meaning that it is thought to be without energy – that is not possible, for thought is order, putting something into order requires energy [576]. Consciousness follows slightly behind most brain activities, because it requires a great expenditure in cooperative activity of neurons. Consciousness about self-and other that is predicated on empathy is a form of inter-pathic communication.

An excitation of the neurons of our retina (which is a part of our brain extended forward) takes – depending on dark-or light adaptation - about sixty milliseconds or more excite our visual cortex, and for the further processing towards conscious perception, take more time [577]. For Waugarman (2007), subtlety of communication lies in, diverting, displacing, and elaborating meanings, expanding an image into a web of associations or condensing a flow of statements into a single focusing insight; in shifting meanings by shifting perspectives or changing the rules for interpretation' [578]. Deecke (2012) argues that there are two evolutionary principles which shaped consciousness, the first of these is the need for data reduction and the second is for a hierarchical organisation of the nervous system. The nature of human evolution has shown that the old pre-existing systems are not 'destroyed' but new layers are built on top with inhibitory controls on the old system, for example within every human there are the rudiments of a fish, reptilian brain and well as mammalian systems [579]. Deecke (2012) states that the most important features of the evolution of the human brain are corticalisation [580]. The human frontal cortex is the organ which controls many subcortical trends and flows to provide episodic memory [581]. Most information is of the spontaneous kind, only the most important information is, therefore stored for longer duration. Only approximately one-per cent of the information flowing through our consciousness, ends up being stored in long-term memory [582].

Unconscious communication, the myriad signals and cues that sensory organ the neocortex filter from the environment, influence behaviour and mood. As Jennifer (2009) argues, "… our unconscious gestures and expressions affect business …" [583] Involuntary signals change our perceptions of risk, reward, and trust. Our unconscious reactions say more than we mean and the expectation is that being non-conscious they are in effect 'truthful' because not changed or altered by conscious cognitions. People communicate clearly and unconsciously through behaviours like nodding, accelerated speech, and fidgeting. However, are not good at knowing how aroused they are or being able to control how efficiently they deliver information. Honest signalling happens without us thinking about it, so we're not much aware of it, it's involuntary even if we try to control it. The unconscious communication actually changes our perception of risk, reward, and trust in social and environmental situations and contexts. Circuits of trust between people circle back and fix in social fabric they make behave more honestly. Charismatic people are really good at sensing and sending off the appropriate signals [584].

THE SEMANTIC WEB AS A SYSTEM OF UNCONSCIOUS COMMUNICATION

The introduction of new technologies are seen as basic extensions of human abilities with telegraph, radio, TV (television) and internet seen as a "planet wide extension of the human nervous system" that transmits and interconnects the thoughts of billions of people around the globe [585]. Email, txts, social networks, and interactive games are bringing about creative modifications of peoples' free-time in all spheres of leisure and business. The world wide-web is a book of the global mind with over one trillion pages – a form of 'exomemory.' Were they to exist, mirror neurons might allow for a deeper passive learning environment [586]. The era of mobile technology is also seen as that of the demise of distance and the disappearance of privacy [587]. But it is a truism that the spoken word is the main 'information technology' used in human society, development of more tools for written language. As Brand (1987) stated of the electronic information age, "[i]nformation Wants To Be Free. Information also wants to be expensive. . . . That tension will not go away" [588]. In conversing through the internet people are redefining personal boundaries – greater and smaller. The invention of 'cloud' computing in which the internet, remote servers and databases that are connected to it become extensions of memory, software and processing power, redefines the concept of the intelligent machine leading to the concept of the semantic web. As Metcalfe's law[14] states, the value of the computing network increases as the square of the number of people who are connected to it [589].

[14] Metcalfe's law concerns the network effects of communication technologies, such as the Internet and the World Wide Web. Mathematically it is represented by the equation $n(n - 1)/2$, representing the number of unique connections in a network of a number of nodes (n). The greater number of nodes, the more valuable the network becomes.

The use of raw data is a fundamental change to the World Wide Web, and open data is a re-orientation of the web. It is suggested that Web 3.0 or the 'semantic web' will no-longer be built on links between documents but forms links between heterogeneous data (which may be qualitative or quantitative) to particular referents – things, places, concepts, people. This is the concept behind the semantic web or Web 3.0., currently the primary unit of the web is the document, but once data is freely available as data, the users of the web or the web itself in responding to algorithms may make links between data. Then the primary unit of the web will become data entities which are real world things, people, places, products in which multiple and heterogeneous data can be linked, or which form complex inter-linkages [590]. To these will be added concepts of free access and usage of data. The development of ontologies is shaped by social networks as well as institutional and interdisciplinary hierarchies finding consensus not easy [591].

It conforms to contemporary web usages in that digital trends indicate the users want (Kristina, 2013) [592].

- Bite size content
- Personalised channels
- Multi-tasking
- Immediacy (hear, new, and now)
- Has an audience
- Capable of being self-actualised
- Visual
- Mobile
- Differentiated
- Transparent

There is thought to be a semantic triangle between communication agents (as real world objects) and concepts (which are the mental image or impression of a real object from the sender's point of view), which may be represented, altered or re-created in the semantic interoperability of an information exchange process [593]. Approaches towards semantic interoperability in the internet are broadly speaking based on shared ontologies which offer a 'common thesaurus' for the information exchange process. In this context an ontology refers to an explicit conceptualisation of the concepts of knowledge domain [594]. The issue of inter-operability and shared ontologies also introduces various secondary problems in consideration for the semantic web, which include scalability, and the desire for the same performance in communication processes as the ontology's size increases. As Chainese et al., (2011), state, "[s]emantic Interoperability is often referred to the ability of two or more computer systems of exchanging information and having the meaning of sent information automatically and correctly interpreted by the receiving system" [595].

Furthermore, there are generally three approaches to meaning creation or semantic inter-operability in communication systems which might contribute to an understanding of a semantic web [596]:

The first approach holds that a communication event is the outcome of the interaction of a quadruple of elements where

- S – is the sender
- R – is the receiver
- M- is the message
- C – is the channel

S and R are agents that exchange a semantic message within a communication channel but the message is not a product of the channel only. For a semantic interoperability all the information as required by the receiving system to interpret meaning correctly whatever the algorithm used (could be unknown to sender).

In definition two (which preserves semantic communication) a communication event for a receiver agent can associate incoming message as being of the same semantic type (referent) of the original message sent by an agent. To achieve S. I. receiver R associates message symbols as same referents intended by S, humans capable of deducing symbols of right referents, but cannot be said of formal systems in same way.

The third definition involves the 'semantic triangle interoperability model' a hybrid approach. This has the components of [597]:

- The communication agents, called semantic machines
- The set of referents of the universe of discourse'
- The set of references for each agent'
- Transmitted message – info object
- The communication process

For information encoding the sender agent composes a message in a format understandable by other semantic systems, tag and referent interrelated by reference mechanism. In information transmission, a sender agent publishes a message (by web pages) on the web, or transmits message to recover agent (for example, by mail). In information acquisition receiver agent performs the message acquisition and in information decoding, receiver binds the information referents by means of references.

Thus according to Pollock (2009) the semantic web will be an upgrade to Web 2.0, use metadata technology for business software, involve open-source data, provide a new generation of artificial intelligence [598]. It will involve rule interchange, personal agents, linked data and a data base – under the legend 'information wants to be free' –for example, to form linked open data in the cloud. Deleuze and Guattri (1987) state it thus, the rhizome "connects any point to another point, and its traits are not necessarily linked to traits of the same nature . . ." [599]. The system is thus decentred, hyper-textual and intratextual. As such reasons for the semantic web include – improving speed of research, protecting national interests, unifying software standards, making software more change-resilient, forming a giant database and providing the gift of open knowledge [600]. However, an environment interacting with a person according to a computer program isn't synonymous with inter-personal communication. But on a spectrum of non-conscious to conscious communication there is also the problem of subliminal communication, which although unconscious nevertheless may involve a subtle sub-conscious register of some king of arousal or communication effect in an individual.

The Problem of Subliminality

The raising of hairs on the forearms or back of the neck, the sudden appearance of good bumps, the reddening of the face in blushing, the spontaneous reddening or darkening of face, ears, neck, and upper chest following unwanted social attention is an example of the involuntary display of emotion. Another is the form of intuitive social judgments of others that people make without conscious thought. The transmission of information between groups without conscious awareness. These are all different examples of unconscious information processing, popularly termed subliminal because it is automatic, not dependent on self-processing, or awareness of the self. Such unconscious actions that are more than physiological process in that they carry communicative intent (or a form of reciprocal intentionality) are nevertheless automatic links between experience and emotional display and are symptomatic of the fact that there is a wide existence of self and between-self socio-cognitive processing which is not dependent on voluntary or subjective activation [601]. As Choi, Gray and Ambady (2004) have noted judgments made on the basis of minimal perceptual glimpses of others may be quite accurate and similarly, impressions formed on minimal information although unreliable can also be valid [602]. Knowledge of the self, one's perception and relation to one's proximity and environment may be both conscious and automatic. Information related to physiology and motor-co-ordination may be an automatic form of unconscious self-processing. This may well be an adaptive survival feature given that unconscious self-processing may contribute to information selection that overcomes the immunity of limitation to conscious cognitive resource [603].

Thus subliminal behaviours and communication results from behavioural effects that are not accessible enough to be reported, which as Geng and Xu (2011) states point to "the existence of physical self-processing unconscious state" [604]. There are thus thought to be three categories of invisible self-stimuli that function as 'primes' for attitude and to sway affective judgment. These are: Sub-conscious or unrecognised self-processing (a conscious process in which one is not aware of the nature of self-relatedness); and unintentional self-processing in which stimuli related to use did not realise a performing self-related processing which was unintentional; and thirdly, a subliminal self-processing in which a psychological or physical self-processing is imagined and in which participants are not aware of embedded information [605]. Furthermore, amongst behaviours and communications that humans are aware of as they process stimuli, people are not necessarily aware of its meaning or self-relatedness or relatedness to others, such communication, perception and reaction may occur in subconscious self-processing or 'unrecognised' self-processing. Unintentional self-processing can also occur from memory absent of the intention involved in the encoding situation [606].

EXPLANATIONS

According to Chartrand, Maddux and Lakin (2004) there are two primary modes of self-representation. These are independent and interdependent. Independent self-construal allows people to think of themselves as autonomous individuals separate from others with unique personal traits. Whereas interdependent self-construal has a context of a larger social world, in which people define themselves in terms of group memberships and relationships with others [607]. Perceiving is the link between one's own behaviour and another's [608]. In this context mimicry performs a function as the social glue that binds people together. Similar facial expressions, speech patterns, moods, emotions builds empathy, rapport and affiliation in mimicry behaviours. Some mimicry is of observable behaviours (whether agentic or consciously motivated) and some automatic behaviour results from spontaneous trait inferences or stereotypical activation on another [609]. Hatfield and Cacioppo (1994) have a notion of emotional contagion mimicking what others do emotionally, so behaviours and actions may not be mimicked for the functional reward they provide but because they are associated with emotions that do provide some reward [610]. The behaviour becomes the language of transferral and experience for the emotion. An experiment by Schacter and Singer (1962) defended a two-factor theory of emotion in which if when people become aroused with no explanation they will label the arousal according to the situation they find themselves in at the time, or if they have sufficient explanation they will be unlikely to look for alternatives [611]. When a person perceives they are engaging in a particular behaviour that is neurologically similar to engaging in it (though they have not) a form of empathy is communicated and self-recognised which is possibly ascribed as being the property of mirror neurons. If a person fails to affiliate with someone, they are more likely to mimic second person. A failure in experience activates goals to rapport with someone new [612].

Consequently, perceptions of behaviour provide information about a wide range of psychological constructs – mood states, dispositional characteristics, social relations, job performance and the understanding that people can be read by emotional expressions [613]. People can express and interpret subtle clues to internal feelings and beliefs however, automatic processes are characterised by unawareness, efficiency, and uncontrollability. Thus Beattie and Ellis (2014) propose a subliminal storehouse which is a "nexus of concerns that include the nature of language and signs, the concept of unconscious, the tugs between subject and agency, and the daily experience of media in late capitalism" [614]. All of which characterise the discrimination without awareness of the subliminal, below the threshold of sensation. In this respect internet 'tags' such as the 'cookies' which learn users web behaviours and preferences and which seek alike information, lead to the transition between Web 2.0 and Web 3.0, the semantic web. However, remote infrastructural sensors that convey information to cloud databases and which are usable on personal electronic devices will greatly enhance the concept of the semantic web. If subliminal communication mediates at some very remote level the communicative exchanges in an information system, this is greatly enhanced in conscious thought by the experience of empathy, which is necessary for interpersonal communication.

Chapter 14

FLOATING SOUNDS

You are sitting in your lounge, and your window overlooks the street below, you have the blinds down because its darkening with evening, you can hear the background noise of the city and its traffic, suddenly, you hear your sister's voice, calling you by name. Your sister lives two hundred miles away, and you haven't left the *Skype* running. You are in a crowded café, standing in line before purchasing your lunch, two women are standing behind you, you suddenly see one of the women pointing and saying under her breath, 'that man's a rat,' or did she say, that? Couldn't she be saying, 'I'll have some of that'! This chapter is about the psychological phenomenology of non-pathological auditory hallucinations. It will define auditory hallucinations on a continuum of normal and abnormal behaviour and describe the main characteristics of them. If you think you have heard voices at some time in your life that you couldn't explain, the good news is that it's perfectly normal and you will be most probably be among the majority of people for whom such a phenomenon is non-pathological, and attributable to a variety of explanations. Julian Jaynes (1976) hypothesized that the origin of human consciousness, or perhaps more importantly, human awareness of their own consciousness, occurred within the last ten millennia, as those voices which were formerly attributed to Gods and supra beings outside a person, became gradually internalised and recognised as emanating from a person's own mind and consciousness. Jaynes (1976) argued there was thus a 'breakdown' of the bicameral mind in human psychological adaptive evolution. Outer voices experienced as separate from a person and 'heard by them' became gradually recognised as the 'inner voices' of their own thinking, particularly as language use became more populous, practised, culturally ingrained and evolved and distant from hand gestures.

It is not that uncommon to 'hear voices' and the majority of people who do have no psychological disorder, this chapter accounts for why and how. The chapter provides definitions of auditory hallucinations; outlines differences between pathological and non-pathological hallucinations, describes a range of hallucinatory benign experiences; suggests possible causes of non-pathological auditory hallucinations; identifies potential perceptual and physiological location of auditory hallucinations; compares differential diagnoses of auditory hallucinations; and briefly discusses psychotherapeutic approaches. 'Hearing voices" is on a continuum of awareness. It is likely that the majority of people may experience some form of perception of auditory hallucination in their lifetime. Most perceptions of auditory voices are benign and non-pathological.

DEFINITIONS AND AETIOLOGY OF AUDITORY HALLUCINATIONS

Sacks (2012) argues that for the majority of the psychological; history of civilization, auditory hallucinations – 'voices' (in a similar way to ocular hallucinations) were attributed to forms of supernatural agency (gods or devils, angels or demons), that such reported experiences could have also been caused by organic or non-organic illness such as hysteria or psychosis but may have been normalised (if kept quiet and private) as religious experiences, for example, which were regarded as a part of 'normal human behaviour' for some groups of people [615]. People may experience ocular or auditory hallucinations for a number of reasons: as a bi-product of a variety of causal factors including but not limited to: physiological disturbance of perceptual and cognitive processing, faulty perceptual cognitions, fatigue, over-arousal, anxiety, grief, depression, obsessive thoughts and possibly organic or non-organic mental illness such as Post Traumatic Stress Disorder (PTSD), caused by over-activity of areas of brain functioning. Lack of insight is sometimes attributed to sufferers of external auditory hallucinations in so much as many may be attributable to internal causes - this is not to confuse them with internal auditory hallucinations which are closer to normal thinking and the abrupt perception of an inner voice [616]. Insight is related to calibration – whether an individual's self-perception is congruent or discordant with a community.

Lack of insight is seen as both a defence mechanism and a neurocognitive deficit as it may be attributable to symptomology of distressing behaviours [617]. There is also a difference between physiological effects on auditory processes and the psychological effects from physiological processes. For example, disturbances in organic (or non-organic) brain physiology may give rise to the psychological perception of auditory hallucination. This implies that the 'location' of the hallucinatory experience is a complex interaction between physiological processes in the brain and the psychological experience of environmental perception.

Perhaps the most common non-pathological auditory hallucination is the 'hearing of a refrain of music' in the brain (it is perceptual and seems to emanate from an exterior source) [618]. Pierre (2010) confirms the view of Sacks, among others, that over the past two centuries, the observation of 'voice hearing' in the 'normal population' and among patients has created a dilemma for physicians who deliberate on diagnoses involving symptoms of psychosis, he makes the point that, "[v]oice hearing is known to have been experienced by various luminary figures throughout history, including philosophers, religious and political leaders, and writers and artists, as well as psychiatrists and clinicians" [619]. The importance of auditory system from a psychotherapy perspective is that it is more closely related to brain functioning that gives rise to self-consciousness than for example the visual system which is more directly related to 'object consciousness.' Seventy-one per cent of subjects in a study by Posey and Losch (1983) identified themselves as having experienced some form of auditory hallucination during consciousness, examples being, hearing one's name called, the voice of religious figures or deceased relatives [620]. Pierre (2010) also identifies the associations between "mood disorder and the artistic temperament and between epilepsy and religious experiences" [621]. Most temporary experiences of auditory hallucinations are mono-symptomatic and do not have accompanying disorder or delusions, they not correlate with significant disturbances in mood (highs or lows) and are non-Schneiderian (result from an inability to self-monitor intentions or actions, or do not have the characteristic of a running commentary, or multiple voices conversing). In this respect temporary auditory hallucinations perceived as fleeting but 'real' perceptions from an outside source can be thought of as ways of "disowning mood-congruent but ego-dystonic thoughts and averting personal responsibility for 'bad' behaviour" – in a sense a problem of misattribution [622]. Pierre (2010) also suggests that such atypical and non-diagnostic symptoms may also be culture specific [623], for example, in micro-cultures of extreme religiosity may 'prime' perceptual experiences of auditory and ocular hallucinations and thus be 'culture-bound' and are thus to be seen as 'idioms of distress.'

A further factor in the experience of non-pathological auditory hallucinations is that they may be symptoms of a kind of 'conversion' disorder which are characterised by insight into the experience, an absence of other symptomology (such as mood or cognitive disorder), they may involve more than one mode of sense (include visual or tactile components) and also have a 'child-like' or naive quality, for example, the experience of a voice from childhood in an admonishing or rewarding state. According to Haddock et al., (1999) three factors are involved in auditory hallucinations: Firstly, emotional characteristic (i.e., negative content and distress); secondly, physical characteristics (e.g., frequency and loudness); and thirdly, cognitive interpretation (beliefs about voices and control) [624]. Hunter (2004) argues that non-pathological hallucinations may be distinguished from pseudo-hallucinations (for example the experience of excess religiosity) which may develop into authentic hallucinations and the idea that such a distinction may depend on whether the experient is conscious of the inauthentic nature of the experience – a voice from outside the head, experienced as an external emanation [625]. In this context Singh et al., (2003), identified the two classificatory factors of the 'reality of hallucinatory perception' and the extent of 'immersion in the hallucination' [626]. He argues that one strand of diagnostic reasoning has its basis in the perceptual model and consequently that auditory hallucinations are problems of perception. This line or argument stems from the contention of Jaspers, that auditory hallucinations may exist alongside real experiences which become perceptually distorted by faulty cognitive processes. As Jaspers (1997) explains, "[h]allucinations proper are actual false perceptions which are not in any way distortions of real perceptions but spring up on their own as something quite new and occur simultaneously with and alongside real perceptions." [627]. Consequently, hallucinations differ from ideas, concepts, and images in so much as they contain false sensory perceptions which may or may not be caused by faulty-processing of cognitions related to perceptual experiences. Langer, Cangas, Perez-Moreno, Carmona, and Gallego (2010) suggest that there exists a continuity between non-pathological experiences of auditory hallucinations and in the normal population and people with a variety of pathological issues [628]. Thus hallucinations can be described as being sensory experiences in a conscious state that take place without external stimulation of a sensory organ but which nevertheless are perceived as true perceptions and not within conscious control. They are thus a disorder of 'self-monitoring.' Some psychotherapists treat hallucinations as a form of pre or conscious thought which is uncontrollable and which are introjected into consciousness and thus symptom of an individual's psychodynamics [629]. Friston (2005) proposes that it is caused by an imbalance between prediction and realised sensory input [630].

Recent technological advances in neuroimaging techniques have led to the ability to represent the symptomology of auditory hallucinations on new scalar models based on neural networks also led to the development of cognitive therapies for auditory hallucinations [631]. The diagnosis differential which is one of the most relevant to the experience of auditory hallucinations is the experience of 'commanding features' – that is the extent to which the perceived hallucination has intentionality and behaviour-specific meaning. For those people who experience auditory hallucinations the evidence points to the perception of commanding aspects amongst between thirty-three per cent to seventy-four per cent of voice hearers [632]. However, the differences between the common experiences of hearing an 'inner voice' and the experience of auditory hallucinations is relatively under-examined. There is a possibility that the two might be confused or that the experience of an inner voice may be externalised and mistakenly attributed to an exterior source. One differing factor is that inner voices are recognisable and controllable and have congruency with other cognitive experiences including the ongoing narrative of consciousness. A recurring theme of experients of auditory hallucinations is that they are less able to correctly identify locations of voice sounds than people who do not have them [633]. Langer, Cangras, Perez-Moreno, Carmona and Gallego (2010) also identify intrusive thoughts and experiential avoidance and anxiety as characteristics of hallucinatory triggers [634].

However, when the experience of a 'voice' does not fit with stream of consciousness thoughts there is the possibility that it represents introjected negative thoughts arising from the subconscious. Thus these experiences may be on a similar spectrum to the automatic thoughts and cognitive distortions that are ameliorated by cognitive behaviour therapy which involves raising awareness of the source of negative thoughts, externalising them as if from an outside source, and rephrasing them in a positive framework. This achieves the reactivation of 'top-down' cognitive processes which involve inhibitory control which raises awareness in non-pathological patients that their voices may stem from inner thought and not from an external source that is connected to them [635]. Furthermore, in some circumstances auditory hallucinations can be experienced as real subjective states but more frequently the experience involves sensing something which is not really existent. The simple relationship between pseudo hallucinations, true perceptions and imagination is given below (See Figure 8).

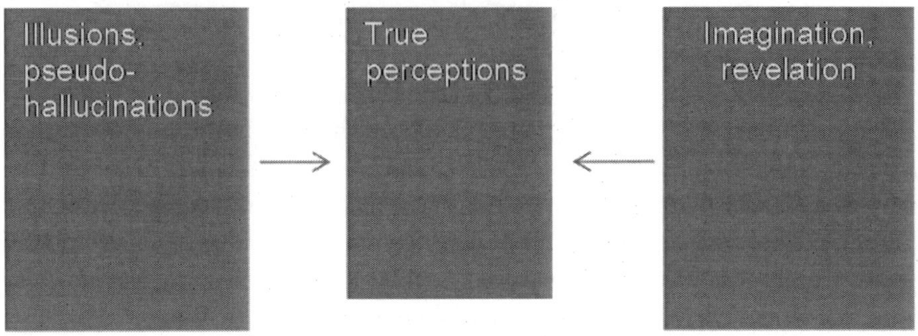

Figure 8. A perceptual continuum.

There has also been a correlation established between the occurrence of auditory hallucinations and post-traumatic stress disorder in so far as their occurrence in context of themes of threat, survivor guilt, and conceptual disorganisation (in the absence of distinct delusions) has been observed [636]. However, the traumatic memories associated with Post Traumatic Stress Disorder (PTSD) have little explanatory effect for the experience of benign auditory hallucinations for example, voices in relief, or 'listening to angels' [637].

Differences between Pathological and Non-Pathological Auditory Hallucinations

Freud gave a thorough definition of non-pathological errors of perception and cognition in his 1914 text *Psychopathology of Everyday Life*, "It [the non-pathological symptom] must not exceed a certain measure [and] [i]t must evince the character of the momentary and temporary disturbance. The same action must have been previously performed more correctly or we must always rely on ourselves to perform it more correctly; if we are corrected by others we must immediately recognize the truth of the correction and the incorrectness of our psychic action." [638]. As Sacks (2012) observes people who experience an auditory hallucination whether of the 'command,' 'musical' or naïve variety may commonly search for an external source – a passerby's voice, a radio, or television (TV) [639] and may either the attribute the source to themselves or remain ambivalent about the causal attribution. Pierre (2010) found that the highest rate of positive response was an occurrence of hearing the doorbell or telephone ringing (sixty-nine to seventy-one per cent), hearing a name called in public (forty-two to sixty-four per cent), or a name called when alone (nine to thirty-eight per cent) none of which could be considered pathologic and many of which are transient experiences [640]. The experience of conversion auditory hallucination is particularly problematic, for example, when an external source is attributed a meaning defined by the perceiver which is unconnected to the meaning at source. Frequently, this can be resolved by acknowledging the perception of an auditory experience and disentangling that experience from the cognitive attribution given to it by the perceiver.

Hugdail (2009) goes further to suggest that if non-pathological people experience 'hearing voices' and show similar brain physiology as potentially pathological experients, there may exist a 'hallucination' factor "independent of disease pathology and diagnostic category" [641]. More than likely such a 'hallucination factor' may have a variety of contributory causes. One possibility is an 'amplificatory effect' which might result in mistaken attribution of cognitive content to unrelated auditory phenomena possibly as the result of over-stimulation or prolonged anxiety which triggers an over-accentuation of bottom-up perceptual processes and a failure of top-down cognitive processes. It is clear that auditory hallucinations experienced as real phenomena can involve single or multiple perceptions, who may be known or unknown, speaking sequentially or simultaneously, in first, second or third person, can be encouraging, insulting, commanding, encouraging or soothing [642]. Hugdail (2009) also reports the experience of hallucination in seventy to eighty per cent or the population and that it may occur on a continuum of benignity, distress and dysfunction [643]. So an hallucinating mind might under (ab)normal conditions be one that is under stress, be that as a consequence of unusual environmental stimuli. It seems that some form of hallucination in response to stress may be the result of a mind that is perspective testing, or it might be due to residual over-excitation of cortical areas, or due to the removal of inhibitory layers of top-down ordering.

Given that people considered not to exhibit features of psychopathology across a normal range of experience do experience auditory hallucinations reasonably often and also that they may be observed clinically across a variety of illnesses not considered symptomatic of psychosis it is difficult to differentiate benign hallucinations from less benign hallucinations [644]. Indeed, a study of college students by Barrett and Etheridge (1992) revealed that up to sixty-four per cent of subjects had experienced 'voice hearing' experiences of perceptions seeming to emanate outside the head with non-definitional causal attributions [645]. However, other studies have found the occurrence rate of auditory hallucinations among the general population to be smaller, approximately seven per cent [646]. Tien (1991), for example reported auditory hallucinations occurring among thirteen per cent of the healthy population in the USA (1991) [647]. Patently this is a phenomenon which occurs outside of Los Angeles, Hollywood and versions of the *X-Men* films.

So 'hearing voices' maybe a normal infrequent occurrence among a healthy populous and Pierre (2010) concluded that the auditory hallucinations of normal people and patients was a similar experience differing only a spectrum of frequency, content, distress and co-occurrence with other symptoms [648]. Clinical and non-clinical diagnoses are also differentiated by such factors as lack of full sensory experience, voice being internal verses external localisation, or lack of insight as distinguishing factors [649]. The variability of these characteristics of auditory hallucinations (partial or absent insight), (soft or loud volume, self or other directed, commanding or rewarding, internal or external localisation), mean that there is little diagnostic predictive value associated with them [650]. The main similarities and differences between pathological and non-pathological auditory hallucinations are then, no significant difference between topographical voice characteristics for hallucinatory experiences (such as frequency, duration, location and loudness) but clinical cases are characterised by increased distress responses to voices, less perceived control and more negativity [651]. An explanation for behaviour that is attributed to your work routine, or the stress you're under, rather than a command from Zeus is more likely to be rationally credible.

Theories of Causation for Auditory Hallucination

Sacks (2012) identifies over-excitation of the primary auditory cortex as the most probable source of auditory hallucinations, in both benign and pathological experients [652]. However, what separates the perception of such voices as 'external' in most experients is not clear but it is likely to be defined in terms of the function of a physiological barrier or inhibition that prevents external attribution or normalises internal monitoring of thoughts [653]. Sacks cites Julian Jaynes provocative book *The Origin of Consciousness in the Breakdown of the Bicameral Mind* (1976) [654] as suggesting that the difference between ancient human and modern human consciousness is that for the former all humans 'heard voices' emanating from within (located in the right brain hemisphere) but understood as external and believed to be communications from outside authorities (such as Gods); but at approximately 1000BC modern human consciousness evolved and the voices became recognisable as 'internal' and as the product of the human mind itself – that is, the hearer's own voice. Some of these internal voices may be the product of 'listening' to both the pre-conscious inclinations and thoughts and the internal qualia, impulses, and sensations of the emotions. As Pinker (1999) states, "[h]ow can our intangible beliefs, desires, images, plans, and goals reflect the world around us and pull the levers by which we, in turn, shape the world? Through the motivation and stimulus that the emotions give us." [655]. Hence voices could emanate from a variety of 'sources' in the body before they coalesce into the 'platform' of conscious planes of cognition. Different impulses, experienced as voices, could be perceived to 'clammer to speak.' Without guided introspection, these voices could be attributable to sources outside the body. Jaynes theory is to some extent corroborated by the attribution of verbal hallucinationsAs Diederen and Sommer (2012) state, " . . . to possession by a spirit in non-Western societies . . ." [656]. It does have significant explanatory power as it accounts for many scientific anomalies in human civilisation such as the origins of religious belief, the development of the interior monologue of consciousness, the evolution of human self-identity, and not least the characteristic of modern human consciousness itself. Thus Jaynes' claim is that the ability to differentiate voices as emanating from within (rather than outside) the mind, is an evolutionary adaptation leading to modern human consciousness.

It remains a task to define how the physiological barrier or inhibitory effect that prevents voices emanating internally from being attributed externally and instead recognised as the experients' own voices in non-pathology symptomology. However, it is clear that auditory hallucinations, 'hearing voices,' are physiologically located in speech areas of the left temporal lobe [657]. It is moreover likely that the kind of 'top-down' physiological inhibitory control is exercised in pre-frontal cortex areas of the brain (known to have undergone evolutionary growth in the era of modern consciousness Jaynes describes) that correct perceptual experiences as internal or externally sourced. Thus there are limitations in the theory that auditory hallucinations arise as lack of inhibition in inner speech of consciousness that are derived from the re-experience of traumatic memories, as this does not account for over-excitation of the absence of external stimulus but it could do in the absence of to-down self-monitoring process – if prefrontal cortical areas are under-stimulated, such control may be absent.

This would also bear a correlation with memory models of auditory hallucination which suggest that they arise from suppressed but automatic and unintentional stored 'experiences' that have been the source of traumatic over-identification. Such conditions might arise in people with divergent mood cycles or under conditions of fragile self-identify formation. However, limitation of both theories are that they cannot account for the perception of an externalised voice outside the hearer that experients of some auditory hallucination report [658]. The latter is supported by the fact that in only ten top twenty per cent is hearing voices about memories or previous experiences [659]. Hence there may in fact be a perceptual cause or cognitive malfunction for the experience of auditory hallucination. If this were so then the speech recognition would be a perceptual error, or disturbance of speech recognition in the left peri-Sylvan region [660]. As perception attention shifts to the exterior voice, a disconnected pre-frontal cortex inhibitory system causes a 'loop' in self-commentary by the inner-voice mis-attributed to the external voice. However, as Hugdail (2009) explains, the left peri-Sylvan region is connected by network neuronal circuitry to anterior regions of the temporal lobe and also with pre-frontal, and dorsally parietal regions [661]. If an auditory hallucination has an experiential perception without a corresponding real event than it is more likely that over-excitation rather than inhibition could be a cause. Hugdail (2009) points to a possible over-excitatory substrate being the neurotransmitter glutamate transmitter which causes neuronal activation in the absence of perception [662]. It is possible that if top-down processes were inhibited in the pre-frontal cortex and glutamate over-excites the peri-Sylvan region (as the bottom-up processes at synaptic level). Amplified by inter-hemispheric transfer across corpus collosum [663]. The 'inner-speech' model is thought to account for auditory hallucinations from a situation of self-monitoring deficit in which a willed intention to think in inner speech is misrecognised, a resulting discharge from the vocabulary of inner speech is recognised which results in a misidentification of intention or action [664]. Thus that part of the cortex which monitors auditory association constructs a transient representation of speech sounds by retrieval from lexical memory. However, in this model the brain could be said to produce auditory hallucinations which are then compared with actual occurrences and dismissed lends support to a theory of over-excitation. Thus from the perspective of subjective agency, auditory hallucinations can be regarded as the passivity phenomena caused by the "... the external misattribution of ... speech if the sense of agency over speech is deficient" [665]. They auditory hallucinations may be related to misattributed triggers for semantic-related false memories. When a form of external stimulus is provided fording an encoding or retrieval, a "critical non-presented associate" (thought or sensory experience) is made more accessible through semantic activation of the neural network [666].

In the auditory-perceptual model of auditory hallucinations, the association cortex in the left plenum temporal is involved in the cognitive process of locating voices in external space (see Figure 9). People are generally better at identifying spatial location of external voices on the right side than on the left as these produce an increased neural response in a dominant left auditory cortex [667]. Because hallucinations are sense-perceptions these must be associated with neuro-transmission in brain regions associated with the location of sounds in external space. Hence, Jaspers (1997) distinguished between pseudo-hallucinations which are located in internal space (and which might arise from misassigned inner speech or interior monologue) and proper hallucinations which have a sense appearance of external space [668]. Due to the shadow created by the head on perceiving sounds from either side binaural cues have intraoral and amplitude differences in resulting from sound perceptions, the neural processing of which occurs in the brainstem. Intensity changes affect synaptic depression which preserves temporal information despite intensity changes. These binaural cues are involved in localizing sounds in the azimuthal plane (an imaginary line drawn through a person's ears) but this is not enough in itself to explain the perception of external sounds in the absence of real sounds. A filtering process is involved that converts sound into spatial representation which is then processed in cognition and attributed a meaning. As Hunter (2004) explains, "... the shape of the ears imposes a spectrotemporal pattern on a sound that is determined by the spatial location of the sound" [669]. This spector-temporal pattern produces sound object characteristics, location in space and cognitive attribution of meaning. In hallucinatory experiences, the process may malfunction in so much as a perceived external cue coincides with a pattern of over-excitation in the left auditory cortex and is attributed a meaning in the pre-frontal inappropriate to the external source due to disruptions in the glutamate neurotransmitter.

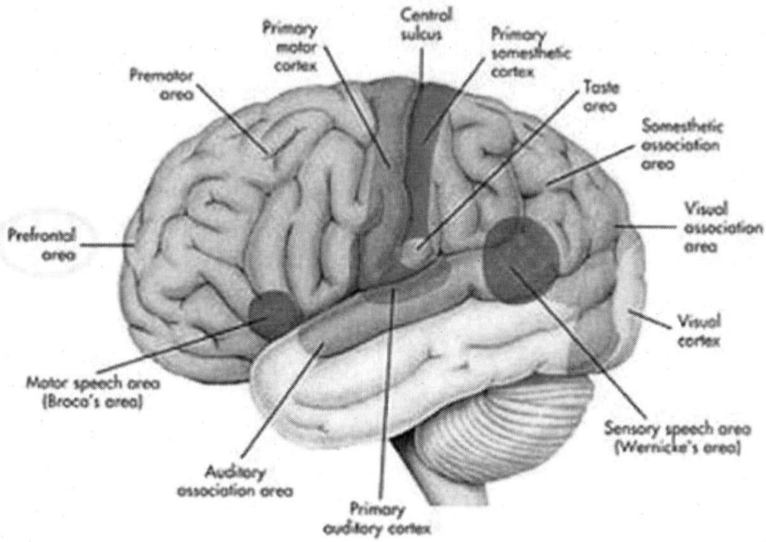

Figure 9. Areas of the brain involved in perception and speech.

However, as Hunter (2004) suggests the plenum temporal region is a likely candidate for disentangling sound location and object characteristics associated with the perception of auditory externality [670]. The expectation-perception model of auditory hallucinations may arise from "abnormalities in the predictive coding which underlies normal perception" and from a deficiency in the cognitive processes which may moderate prediction error in thought processes [671]. Nazimek, Hunter and Woodruff (2012) suggest that there are five possible reasons for this: firstly, hyperactivity in the auditory cortex; secondly, abnormal synapse connectivity and neurotransmitter dysfunction; thirdly, abnormality in the physiological behaviour of the auditory and association cortex; fourthly, impaired transmission of perceptual input; fifthly, impaired modulation of the thalamus region by the pre-frontal cortex (related to top-down and bottom-up processing of external input) [672]. The latter cause is supported by a meta-analysis of neuro-imaging studies on patients with clinical symptoms of auditory hallucinations by Kompus, Westerhausen and Hugdail (2011) who suggest that ". . . increased activation in the absence of an external stimulus, and decreased activation in the presence of an external auditory stimulus in the left primary auditory cortex and in the right rostral prefrontal cortex" which is characterised as a somewhat 'paradoxical' physiological symptom [673].

As Pierre (2010) states auditory hallucinations are frequently found to occur in people who suffer from grief but in this state are perceived of as helpful and comforting [674]. However, the persecutory side of auditory hallucinations are also liked to abuse. One study of people with auditory hallucinations also found that experients had suffered from childhood abuse ranging from thirty-eight perecent to seventy-three per cent of positives [675]. However, it is difficult to determine a causal link as such experiences may be common in people presenting symptomatology of other cognitive disorders. There are also 'straightforward' but unusual occurrences of people reporting hearing voices that are attributed to actual radio transmissions through dental, fillings, hearing aids, and metal jewellery [676]. However, the effects and frequencies of these errant transmission receptions are localised and specific and bear no causal relationship with either the external source or cognitive processes in the brain. A further hypothesis is that auditory hallucinations may have a relationship to certain kinds of cognitive dissociations experienced with speech impairment or aphasia given the shared location with the Broca and Wernicke areas of the cortex associated with speech. Cognitive impairment in these locations share ambiguation with " . . . residues of impressions which reach the brain via the . . . auditory nerves, or which originate as sensations of innervations or of perceptions of movements carried out in the act of speaking." [677]. Functional gaps may be associated with spaces between interconnected white fibre tracts in these cortical areas which transmit 'glosso-kinaesthetic impressions.' It is possible that symptomologies of auditory hallucinations and mild transcortical sensory aphasia exhibiting paraphasia are related.

Differential Diagnosis Spectrum and Psychotherapy in Auditory Hallucinations

Langer, Cangras, Perez-Moreno, Carmona and Gallego (2010) view depression as an important factor in the potential experience of auditory hallucinations due to anxiety and phobia arising from experiential avoidance, which is defined (not as a matter of retreat from reality) but rather as a reluctance to maintain an interlocutory perspective on certain private events (thoughts and emotions that have helped to construct memories). As they explain, the top-down self-regulatory awareness produces ". . . self-focused attention, attentional bias, activation of dysfunctional beliefs, ruminative processing" which lead to mistaken identificatory perceptions of self-knowledge [678]. One therapeutic process which is designed to counteract auditory hallucination is 'acceptance and commitment therapy' which aims to help enable a person to disentangle their symptoms and become more attuned to values which are more important in their life, this modifies the interpretation and function of symptoms of auditory hallucination and enables the experient to assign diminished credibility to the experience [679]. The factors that make auditory hallucinations clinically relevant are their frequency, the extent to which they preoccupy the person, the extent of the person's insight into them, the extent to which the person is influenced or controlled, and the pervasiveness or omnipotence of the symptoms within the person's conscious experience [680]. As Lera et al., (2011) suggest strategies for treatment include operant conditioning models related to mutual support and self-selective reinforcement, psycho-education, sense-making, coping, and empowerment therapies [681].

As Pierre (2010) states auditory hallucinations are commonly found among experients without any other clinical symptomatology as well as in clinical disorders in which other symptomatology is present. This might imply that auditory hallucinations are not as symptomatology of pathology as they are commonly believed to be, or that experiences of an auditory hallucination – not which are in fact less so, are often mischaracterised [682]. As Pixley (2012) point out, in terms of therapeutic engagement the determination of whether a single occurrence of auditory mis-assignation is hallucinatory or not is not necessarily helpful in determining the difference between non-pathological and pathological features, rather the quality of the person's state of mind or belief in the reality or otherwise of the experience is more helpful in diagnosis [683]. It is therefore the role of the therapist to engage the client in interlocutory conversation and reverse the condition of experiential avoidance into a state of being in which given private events (thoughts and emotions that have formed traumatic memories) can be integrated within current consciousness and desired personality constructs. However, time and the removal of a triggering stressor (whether actual or imaginary) is also thought to elevate symptom logy by enabling the person to reintegrate the source experiential avoidant condition and reintegrate the misattributed voice to recognising it as their own [684].

There is also a vast difference between the hearing of hallucinatory voices of command and compliance with commanding voices in symptomatology. Command voices in themselves are of 'limited risk compared to non-clinical population' but compliance with command voices and engagement in dangerous behaviours are of significantly greater risk [685]. Higher compliance to commanding voices may be related to feelings of social inferiority and low-power among sufferers of auditory hallucinations. Compliance is defined as an agreement to the perceived will of others with non-recognition of the underlying meaning or its moral and ethical consequences [686]. A further factor that Pixley (2012) observes in the diagnosis and treatment of auditory hallucinations is the recognition of an 'objective cultural reality' that exists alongside an 'objective reality' confirmed by reasonable people and professionals [687]. The implication being that the occurrence of auditory hallucinations may have cultural specific components.

For any experient, the location of an auditory hallucinatory experience (frequently a disorientating experience) is a complex process involving physiological processes in the brain (the auditory and pre-frontal cortex) and the perceptual processing of environmental cues. Auditory hallucinations are relatively uncommon but nevertheless widely occurring phenomenon amongst a general healthy population. The experience of auditory hallucination is on a continuum between non-pathological and pathological symptomatology determined by the quality and frequency of occurrence. It is likely that suffers of non-pathological auditory hallucinations can remediate the phenomena by the retraining of top-down cognitive processes of auditory attribution. So if you hear a voice you don't recognise, tell yourself, quiet, I'm trying to concentrate!

Chapter 15

TOUCH WOOD

Human beings are fragile entities. There is a just a thin membrane of skin some two-to-three millimetres in width separating us from the environment. We are vulnerable to physical damage from the natural world and the civilised world we create. Moreover, the stresses and strains of modern life can take toll on us psychologically. The one defining characteristic needed by humanity in the twenty-first century, physically and psychologically is resilience. We need to know how to look after ourselves physically and psychological, both as individuals and as members of the societies we create. For the most part we may be successful at this, on average, the human lifespan is extending. A person born today can expect to live ten per cent longer than someone born a hundred years ago. But we need to be resilient to guard against the shocks that life can through at us – relocation, migration, natural disaster, climate change, all these need thinking about and preparing for, beyond the folk wisdom that our parents and grandparents may have taught us. Resilience is defined by Norris et al., (2008) as: "[a] process linking a set of adaptive capacities to a positive trajectory of functioning and adaptation after a disturbance" [688]. Longstaff and Yang (2008) observe that resilience is the "capacity of a system to absorb disturbance, undergo change" yet still function in the same way [689]. Thus resilience is the ability of a system to recover from setback with no major change in its form or functioning. Norris et al., (2008) also define five dimensions along which resilience may be enhanced by 'networked adaptive capacities.' These are: firstly, developing economic resources – reducing inequities; secondly, enhancing social capital through minimising harm through damage mitigation; thirdly, adaption and utilisation of pre-existing organisational networks; fourthly, protecting naturally occurring social supports; fifthly, community planning [690]. Consequently, central planners engaged in community rehabilitation strategies should look to strengthen these community functions in post-catastrophe environments. Hence resilience may refer to adaptive governance "across spatial scales (local to national) and between sectors (government, community)" [691]; it may refer to knowledge, communication and social learning, and maintaining a positive outlook and drawing on social networks for support. Surviving a natural catastrophe and rebuilding a community after it, requires an investment in human capital that has the quality of being able to be self-sustaining by experiential resilience.

The basic idea of security is that people should be secure in their daily lives, whether or not they feel secure. It is based on an idea of security that is both within and outside the dominant state-centric argument of security [692]. There is a narrow definition of human security that argues that it constitutes the threat of political violence to the people by the state or any other authority. In terms of human development, it involves 'freedom from fear,' 'freedom from want' and 'freedom from threat and danger' [693]. When basic needs are met and fears allayed, people potentially experience feelings such as security [694].

Peterson and Park (2006) also make the point that human security – either of individuals or groups – must not be confused with the removal of sources of threat. The argument for security is based on the idea that the fulfilment of human potential is only possible in a secure environment and forming a secure environment may only be possible under secure conditions. Danger reduction, whether it is man-made or natural, may be accomplished by avoidance, protection by nature, protection by human design - but the reduction is only a beginning. It stands to reason that physical safety is a necessary but not a sufficient condition for security because psychological security may be independent from conditions of physical safety. Even in circumstances of safety from external threat, if people are not flourishing, then the psychological well-being and the good life has not been attained [695]. The components of psychological security may involve many of those characteristics that one might associate with a positive psychology and the experience of emotionally fulfilling traits. These include: Firstly, positive subjective experiences (happiness, pleasure, and fulfilment); secondly, positive individual traits (strengths of character, attitudes, and values); and thirdly, positive institutions (families, schools, businesses, communities, societies). Values for a secure society may include mutual care, trust and norms for fairness along with positive duties and/or negative injunctions.

Human security, both physical and psychological, may also derive from attachment to other people, usually biological or adopted parents. As such, attachment is a behavioural system that emerged as an adaptive feature over mammalian evolution because it increased the likelihood of survival during infancy. Thus in infancy as in other periods of life, attachment is important for protection and support given that people and other mammals are born without all the capabilities for survival (which must be learned from cohabitation) [696]. The security derived from attachment is especially needed during the early years of life but it is also needed over one's entire lifespan. Bowlby (1988) observed the adult person's tendency to seek proximity and support when distressed [697]. In humans and mammals, the attachment system and the behavioural system are shaped by experiences with attachment figures and memories of experiences respectively [698]. Perception, experiences and memories of experiences combine to form working models which optimally result in 'stable' attachment styles or orientations, which in turn comprise a person's characteristic pattern of emotions and behaviours [699]. The stable experience of human connectedness is a form of security behaviour. Individual differences in attachment anxiety and avoidance reflect a person's internal modelling of experiences, thoughts and behaviours and the way in which he or she copes with distress. For people low on these dimensions, well developed models of comfort-providing attachment figures create a sense of attachment security, and positive self-regard and confidence in proximity-seeking strategies [700]. People who are high on attachment anxiety and avoidance dimensions, possess unavailable working models and unreliable, frustrating or rejecting attachment figures [701].

Insecure people rely on what Mikulincer and Shaver (2003) call secondary attachment strategies – involving either deactivating or hyper-activating – a way of coping with insecurities and worries [702]. Adult attachment orientations are implicated in coping with stress, distress, regulation, and adjustment [703]. People who score low on attachment anxiety and avoidance, hold optimistic attitudes about coping with life's hardships, possess strong self-efficacy with respect to coping with stress, appraise life events in an optimistic way and rely on support seeking as a means of coping with threats, thus are emotionally stable when coping with stress. According to Baldwin (2007), security may result from a self-other working model or relational schema that is activated by a prime, which results in an arousal state producing feelings of warmth and safeness and an expectancy that others might be warm and accepting and trustworthy also [704]. Insecurity is rooted in a feeling of separation from the comforting and protecting figure of childhood – contact with whom could produce emotional security. Security thereafter becomes associated with a sense of social connection, or the shared boundaries with another, whilst maintaining an individual sense of self [705] As Baldwin (2007) states, ". . . security is engendered by warm soothing connections with others, and taken away by anxious, conflictual and rejecting interactions" [706]. Acceptance is formed through a continual co-exchange with another stable figure who can be counted on and not taken away, however, social acceptance is unpredictable and may not be contingent on being successful or be constantly available through good times and bad [707].

As Harries (2008) states, people may put their 'ontological security' before their physical security, therefore believing from the beginning that environments are innately safe, that nature is a positive moral force and that society is a competent provider and protector. This leads to the idea that day-to-day, people don't really need to protect themselves – thus defending the feelings associated with these beliefs and perceptions and representations of 'home,' 'nature' and 'society' are confirmed by avoiding behaviours that might challenge them [708]. An ontologically secure person is free from existential doubts and able to live life without threat to time, space and identity, protected by representational barriers through which information about the world is filtered [709]. People more typically feel closer to others who are warm and accepting – one's propensity toward growth, connection intimacy and exploration is not reducible to anxiety reduction nor explained by the soothing figures in a nurturing role of adults providing a responsive environment. A positive environment might include attachments to people who are accepting, loving, non-controlling, non-threatening and supportive [710].

Chapter 16

A Horizon

The human species is currently evolving. That is a basic fact of the environmental evolutionary process. Following the Hardy-Weinberg equilibrium,[15] evolutionary change occurs so long as any of the processes of mutation, non- random mating (including sexual selection), gene flow, genetic drift or natural selection, take place within a genetic population [711]. But separately from the evolutionary process, the human species is able to adapt and modify its environment within the ecosystems of Earth to unprecedented degree, even to the extent in being able to intervene in the processes of the creation of life itself. In the not too distant future we can modify evolution from our own design. We can 'design' our genes as an extension of knowledge about the genome and medical care.

In the approximately six million years of human evolution, from ape-like hominids to homo sapiens, and the use of tools approximately two-and-a-half-million years ago, we have evolved thumbs, bi-feudalism, and a larger brain. As Harare (2014) observes, "[l]ike a government diverting money from defence to education, humans diverted energy from biceps to neutrons. A chimpanzee can't win an argument with a Homo Sapiens but the ape can rip a man apart like a rage doll" [712]. We have not been 'modern' for very long, really only the last two hundred years since the industrial revolution, and have technological advances such that we can build civilisations of concrete, steel and wood. We have come a long way since the basic tool of the stone-axe some fifty thousand years ago and the development of agriculture and farming techniques from the Holocene era some eleven thousand years ago. As we can

[15] The Hardy-Weinberg principle holds that in the absence of extraneous influences, an allele and genotype frequency in population will remain constant from generation to generation.

plant crops and modify the landscape for agriculture on large scales, we can travel in speeds in cars, boats and aircraft that far exceed our own unassisted ability. We can create energy from nuclear, solar and wind technology. In the future our nuclear technology may be replaced by fusion technology – even more powerful and sustainable.[16] The human population has doubled in size over the last fifty years but in what ways are humans evolving? As Harare (2014) states, "[h]umankind ascended to the top so quickly that the ecosystem was not given time to adjust." [713]. Our brain is a remarkable instrument. As Mossman (2014) reflects, "[t]he human brain consumes about 20 per cent of the available energy intake but constitutes only about 2 per cent of body weight and has an energy flow density 150 000 times that of a typical star" [714]. The emergent properties of the brain such as language and consciousness, seem to have non-linear capacity, they produce more than the sum of the parts, and indeed coalesce meaning itself – they very essence of our notion of 'being.' What characteristics and traits might we evolve in the near-distant future? Firstly, in the last one hundred thousand years of human development our teeth have grown smaller and halved in size. Thirty-five per cent of people today are born without wisdom teeth, and many others with just one or two [715]. Hence when a trait is no longer necessary evolution saves energy on its development by eliminating it. People of the future may have no wisdom teeth. Similarly, our little toe is increasingly of not much use to us. Where formally toes were used for 'gripping' and balance' now they are housed in shoes in which the smallest toe becomes a 'freeloader' on a ride to the next shoe step – there may be pressure to naturally select against it [716].

We have an unprecedented ability to travel over vast distances, our world is globally connected, and people of all culture trade and exchange, converse and inter-breed far more than they have at any other time. Hence race differences are also disappearing. In a global world in which genes are flowing between populations, ethnic differences are disappearing. People of the future may become monocultural with a blend of harmonised racial features [717]. These members of a single ubiquitous ethnic group, may also lose the

[16] The leading development for Fusion technology in the world today is an international engineering megaproject called ITER (International Thermonuclear Experimental Rector) which is an consortium led by thirty-five participating nations. It is a large-scale experiment for magnetic confinement plasma physics, in effect to re-create the kinds of fusion that occur in stars like the Sun. It is an attempt to bridge the gap between experimental plasma physics to full-scale electricity producing power stations. ITER is based on a tokomak nuclear fusion reactor built by the Cadarache facility in southern France. ITER is designed to produce 500 megawatts of power from a fifty megawatts operation source. Nuclear fusion is seen as a 'holy grail' alternative to nuclear fission, with potentially less radio-active by-product.

distinguishing features of their original ethnicities. Our clothing air conditioning and heating systems, have meant that we are making the insulating properties of body hair obsolete. People of the future may have less body hair [718]. Three of the largest sources of disease and mortality in the developed world are cancer, diabetes and heart disease. Hence there will be pressures to select against these genetic propensities and furthermore, solutions for finding advanced treatments for these diseases will come from tailoring drugs to knowledge of individual genomes.

Since the agricultural evolution lactose intolerance has greatly decreased amongst human populations. Northern Europe which is traditionally more dependent on goat and bovine milk has shown this tendency to the greatest extent, though it is changing in the Americas and Asia also. Because of our increased reliance on a vast 'exoskeleton' of automated machinery, people of the future may become physically weaker. It is already known that homo sapiens is physically weaker than other close species of hominids such as Neanderthal population of the pre-ice age. Machines and technology increasingly do the physical work for us and because of this reliance we are weaker than are distant ancestors.

Medical technology has wrought great benefits but we have weakened immune systems, we are more susceptible to pathogens. While medical technology and antibiotics has greatly enhanced our health and life expectancy our immune systems need to do less work to keep us healthy, hence there is less pressure for selection. Increasingly as synthetic Supplements regulate our hormones, the process which create hormones would become less important for survival, and in tens of thousands of years likely to evolve to point where hormones no longer created organically within body, external aids to our internal functions making them obsolete. People of the future will be taller. Human height has been increasing over the last two centuries, because of advances in nutrition. People are on average ten centimetres taller than they were in the middle of the nineteenth century. Furthermore, the developed world has complex systems of technology – safe buildings, access to food (supermarkets) and healthcare, and leisure and exercise. We consume nutrition derived from agricultural products not hunter-gathering. Hence the 'flight or fight' response of our amygdala which 'readies' the human being for response through adrenaline and stimulating anxiety through the limbic system will no longer be so actively selected for. Our emotional lives may become less prone to large extremes.

Our information and computer technology (ICT) enhances our intellectual labour. With GPS satellites and silicon digital storage power, we are no longer reliant on our own possibility innate senses of direction or memory. Information storage has meant that we have less reliance on our own memory. This may be liberatory or it may be folly. People of the future may live for the present far more than those in the past. What if we in the ‘collective’ sense, with ‘our’ incredible technological prowess, forget the lessons that history has taught us – the insufferable harm of warfare, the intolerable damages to the biosphere, for example? For now, the good news is that people are also living longer than in the past, and women are having children at later ages, the body clock may be slowing down as lifespans become greater [719]. Finally, human beings may ‘force’ evolution upon themselves through the use of genetic modification, through gene section, ‘designer babies’ that drive negative traits into extinction. The human species that exists today is but a survivor of many different environmental changes, including adaptation to changing ecosystems and climate conditions. We are a ‘product’ of an evolutionary process of natural selection with our environment though we can now modify not only our ecosystems to an unprecedented extent, but also our own genetic make-up to some extent. Humans are robust as a species, though our brains and organs are increasingly delicate. Potentially, humanity is losing its robustness to the world we create after our own kind, can we trust ourselves and evolution enough, to ‘leave our species alone?’

REFERENCES

[1] Dawkins, R. (2006). *The Blind Watchmaker.* London: Penguin: 116.

[2] Dawkins, *ibid.,* 115-116.

[3] Simon, J. (2002). Human Dignity as a Regulative Instrument for Human Genome Research. (Pp. 35-45). In *Ethics and Law in Biological Research*, Edited by Cossimo M. Mazzoni, Hague, Netherlands: Kluwer Law International: 39.

[4] Simon, *ibid.,* 39.

[5] de Melo-Martin, I. (2005). Firing up the nature/nurture controversy: bioethics and genetic determinism. *Journal of Medical Ethics,* 31 (9): 526.

[6] Mazzoni, C. M. (2002). Editor., *Ethics and Law in Biological Research.* Hague, Netherlands: Kluwer Law International: 128-129.

[7] Rodota, S. (2002). Making the Right Use of Law: The Cloning Dilemma. (Pp. 153-161). In *Ethics and Law in Biological Research*, edited by Cossimo M. Mazzoni, Hague, Netherlands: Kluwer Law International: 153.

[8] de Melo-Martin, *ibid.,* 526-527.

[9] Council of Europe. European Treaty Series No. 164. Convention for the Protection of Human Rights and Dignity of the Human Being with regard to the application of Biology and Medicine: *Convention on Human Rights and Biomedicine.* Retrieved from: https://rm.coe.int/CoERMPublicCommonSearchServices/DisplayDCTMContent?documentId=090000168007cf98.

[10] Rawls, J. (1971). *A Theory of Justice.* Cambridge, M.A: Harvard University Press.

[11] Dawkins, *ibid.,* 125.

[12] *Genetic Information Nondiscrimination Act* of 2008. Pub.L. 110-33. Http://legislink.org/us/pl-110-233) 122 Stat. 881.

[13] Busnelli, F. D. (2002). The Problem of Reproductive Cloning. (Pp. 125-133). In *Ethics and Law in Biological Research*, edited by Cossimo M. Mazzoni, Hague, Netherlands: Kluwer Law International: 129.

[14] Winston, R. *Human Instinct*. London: Transworld: 21-23.

[15] Winston, *ibid.,* 22.

[16] Dawkins, *ibid.,* 125.

[17] de Melo-Martin, *ibid.,* 526.

[18] Harare, Y. N. (2014). *Sapiens: A brief history of mankind.* London: Harvill Secker: 13.

[19] Ekstig, B. (2015). Complexity, Natural Selection and the Evolution of Life and Humans. *Foundations of Science,* 20, (2): 186.

[20] Silvers, S. (2011). Methodological and Moral Muddles in Evolutionary Psychology. *Homo Oeconomicus,* 28 (3): 298.

[21] Mossman, K. (2014). *Complexity Paradox: The More Answers We Find, the More Questions We Have.* Oxford: Oxford University Press: 181.

[22] Mossman, *ibid,* 182.

[23] Crawford, C. The Theory of Evolution in the study of Human behaviour: An Introduction and Overview. (Pp. 3- 41). In Charles Crawford and Dennis L. Krebs, (2002). *Handbook of evolutionary Psychology: Ideas, Issues and Applications.* London: Taylor and Francis: 3.

[24] Diamond, J. (2012). *The World Until Yesterday.* London: Penguin: 7.

[25] Crawford, *ibid,* 5-7.

[26] Wilke, A. and Mata, R. (2012). *Cognitive Bias.* (Pp. 531-535). In V. S. Ramachandran (Edited). *The Encyclopedia of Human Behavior.* Volume 1. Second Edition Academic Press: 533.

[27] Crawford, *ibid.,* 35.

[28] Oakley, T. H., and Speiser, D. I. (2015). How Complexity Originates: The Evolution of Animal Eyes. *Annual Review Of Ecology, Evolution and Systematics*, *46*237-260. doi:10.1146/annurev-ecolsys-110512-135907: 239.

[29] Hughes, G. M., Teeling, E. C., and Higgins, D. G. (2014). Loss of Olfactory Receptor Function in Hominin Evolution. *PLOS One*, 9 (1): 1-6.

[30] Pinker, S. (1999). *How the Mind Works.* London: Penguin.

[31] Wynne-Edwards, V.C. (1962). *Animal Dispersion in Relation to Social Behavior*. Oliver and Boyd, London.

[32] *Darwin, C. (1872), The Origin of Species by Means of Natural Selection, or the Preservation of Favoured Races in the Struggle for Life, (6th ed.) London: John Murray.*

[33] Trivers, R. L. (1971). The Evolution of Reciprocal Altruism. *Quarterly Review of Biology,* 46 (1): 35-57.

[34] Crawford, *ibid.,* 12.

[35] Crawford, *ibid.,* 27.

[36] Crawford, *ibid.,* 23.

[37] Kenrick, D. T., Sadalla, E. K. and Keefe, R. C. *Evolutionary Cognitive Psychology: The Missing Heart of Modern Cognitive Science.* (Pp. 485-514). In Charles Crawford and Dennis L. Krebs, (2002). *Handbook of Evolutionary Psychology: Ideas, Issues and Applications.* London: Taylor and Francis.

[38] Pinker, *ibid.*

[39] Goldfinch, A. (2015). Subverting the Paradigm. (Pp. 29-61). *Rethinking Evolutionary Psychology.* Basingstoke: Palgrave Macmillan.

[40] Wynne-Edwards, *ibid.*

[41] Trivers, *ibid.*

[42] Goldfinch, *ibid.,* 45.

[43] Goldfinch, *ibid.,* 45.

[44] Merkel-Bobrov, N., Gilbert, S. L., Evans, P. D., Vallender, E. J. , Anderson, J. R., Hudson, R. R., Tishkoff, S. A., Lahn, B. T. (2005). Ongoing adaptive evolution of ASPM, a brain size determinant of Homo sapiens, *Science* 309 (5741): 1720-2.

[45] Burger, J., Kirchner, M., Bramanti, B., Haak, W., and Thomas, M. G. (2007). Absence of the lactase-persistence-associated allele in early Neolithic Europeans. *Proceedings of the National Academy of Sciences USA* 104(10):3736-3741.

[46] Ehrenpreis, E. D. (2007). Lactose intolerance: Definition, Symptoms and Treatment. International foundation for Functional Gastrointestinal Disorders.

[47] Pinker, *ibid.,* 42.

[48] Benitez, Burraco, A., and Boeckx, C. (2015). Possible functional links among brain- and skull-related genes selected in modern humans. *Frontiers in Psychology,* 6 (794). Doi: 10.3389/fpsyg.2015.00794: 12.

[49] Hampton, S. J. (2009). Brains, Minds and Consciousness. *Essential Evolutionary Psychology.* Thousand Oaks: SAGE Publication.

[50] Kaufman, D. (2001). Comparisons and the case for interaction among Neanderthals and early modern humans in the Levant. *Oxford Journal of Archaeology,* 20: 234.

[51] Sankararaman S, Patterson N, Li H, Pääbo S, and Reich D. (2012). The Date of Interbreeding Between Neanderthals and Modern Humans. *PLOS Genet.* 8 (10):e1002947. doi: 10.1371/journal.pgen.1002947. Epub 2012 Oct 4: 1.

[52] Villa, P. and Roebroeks, W. (2014). Neanderthal Demise: An Archaeological Analysis of the Modern Human Superiority Complex. *PLOS One* 9 (4): 2.

[53] Gilligan, I. (2007). Neanderthal extinction and modern human behaviour: the role of climate change and clothing. *World Archaeology,* 39 (4): 499.

[54] Gilligan, *ibid.,* 499.

[55] Gilligan, *ibid.,*509.

[56] Gilligan, *ibid.,*509.

[57] Hampton, S. J. (2009). Brains, Minds and Consciousness. *Essential Evolutionary Psychology.* Thousand Oaks: SAGE Publication: 110.

[58] Hampton, *ibid.,* 111.

[59] Hampton, *ibid.,* 111.

[60] Hampton, *ibid.,* 112.

[61] Dillman, R. (n.d.) *The Communication Environment.* HFCL Tutorial. Retrieved from: http://www.rdillman.com/HFCL/TUTOR/ComEnv/ComEnv1.html. para 1.

[62] Pinker, *ibid.,* 272.

[63] Ord, T. J., Charles, G. K. M., and Hofer, R. K. (2011). The evolution of alternative adaptive strategies for effective communication in noisy environments, *The American Nationalist,* 177 (1): 54-64.

[64] Ord et al., *ibid.*, 55.

[65] Johnsen, S., and Lohmann, K. J. (2008). Magnetoreception in animals. *Physics Today,* 61 (3): 29.

[66] Belova, N.A. and Acosta-Avalos, D. (2015). The Effect of Extremely Low Frequency Alternating Magnetic Field on the Behaviour of Animals in the Presence of the Geomagnetic Field. *Journal of Biophysics,* 1-8. DOI: 10.1155/2015/423838.: 5.

[67] Harare, *ibid.,* 10.

[68] Broditsky, L. (2011 Feb). How Language Shapes Thought. *Scientific American.com,* 43.

[69] Broditsky, *ibid.,* 45.

[70] Germonprez, M., and Zigurs, I. (2009). Task technology, and tailoring in communicative action: An in-depth analysis of group communication. *Information and Organization.* 19: 23.
[71] Chellali, A., Dumas, C., and Milleville-Pennel, I. (2011). Influences of haptic communication on a shred manual task. *Interacting with Computers.* 23: 318.
[72] Rasmussen, J. (1983). Skills, rules, knowledge: signals, signs and symbols and other distinctions in human performance models.' *IEEE Transactions on SMC,* 13 (3): 221.
[73] Najafi, M., and Shariff, M. K. B. M. (2011). The Concept of Place and sense of Place in Architectural Studies. *International Journal of Human and Social Sciences*, 6 (3): 187.
[74] Najafi and Shariff, *ibid.,* 188.
[75] Najafi and Shariff, *ibid.,* 188.
[76] Rogen, R. O'Connorb, M., and Horwitza, P. (2005). Nowhere to hide: Awareness and perceptions of environmental change, and their influence on relationships with place. *Journal of Environmental Psychology,* 25: 189.
[77] Najafi and Shariff, *ibid.,* 188.
[78] Simpson, G. G. (1964). *This view of life.* New York: Harcourt, Brace and World: 106.
[79] Gibson, J. J. (1977). *The Theory of Affordances.* In R. E. Shaw and J. Bransford (Eds.), *Perceiving, Acting, and Knowing.* Hillsdale, NJ: Lawrence Erlbaum Associates.
[80] Stewart, A. E. (2007). Individual Psychology and Environmental psychology. *The Journal of Individual Psychology,* 63 (1): 67.
[81] Adler, A. (1938). *Social Interest: A Challenge to Mankind.* Translated by J. Linton and R. Vaughan. London: Faber and Faber.
[82] Petrusz, S. C., and Turvey, M. T. (2010). On the Distinctive Features of ecological Laws. *Ecological Psychology*, 22: 46.
[83] Petrusz and Turvey, *ibid.,* 47.
[84] Petrusz and Turvey, *ibid.,* 47.
[85] Francovich, C. (2010). An Interpretation of the Continuous Adaptation of the Self/Environment Process. *The International Journal of Interdisciplinary Social Sciences*, 5: 311.
[86] Petrusz and Turvey, *ibid.,* 64.
[87] Stewart, *ibid.,* 68-69.
[88] Stewart, *ibid.,* 69.
[89] Stewart, *ibid.,* 69.

[90] Stewart, *ibid.,* 78.
[91] Binnie-Dawson, J. L. M. (1982). A Bio-Social Approach to Environmental Psychology and Problems of Stress. *International Journal of Psychology*. 17: 397.
[92] Binnie-Dawson, *ibid.,* 397.
[93] Lakoff, G. (2011). *Why Environmental Understanding or 'Framing' Matters: An Evaluation of the EcoAmerica Summary Report.* Retrieved 7 November 2011 from: http://not4p.wordpress.com/resources-and-readings/why-environmental-understanding-or-framing-matters-an-evaluation-of-the-ecoamerica-summary-report/para. 5.
[94] Stewart, *ibid.* 71.
[95] Stewart, *ibid.,* 70.
[96] Stewart, *ibid.,* 71
[97] Stewart, *ibid.,* 72.
[98] Jordan, S. (2009). Forward-Looking Aspects of Perception-Action coupling as a Basis for Embodied Communication. *Discourse Processes*. 46: 130.
[99] Jordan, *ibid.,* 130.
[100] Jordan, *ibid.,* 131.
[101] Jordan, *ibid.,* 131.
[102] Francovich, *ibid.,* 313.
[103] Jordan, *ibid.,* 137.
[104] Jordan, *ibid.,* 139.
[105] Craik, K. H. (1977). 'Multiple scientific paradigms in environmental psychology. *International Journal of Psychology*. 12 (2): 149.
[106] Francovich, *ibid.,* 315.
[107] Adams, W. A. (2007). The Primacy of Interrelating: Practicing Ecological Psychology with Buber, Levinas and Merleau-Ponty. *Journal of Phenomenological Psychology*, 38: 24.
[108] Francovich, *ibid.,* 314.
[109] Adams, *ibid.,* 50.
[110] Adams, *ibid.,* 51.
[111] Hoffmeyer, J. (2008). *Biosemiotics. An Examination into the Signs of Life and the Life Signs.* Scranton University Press.
[112] Bertolotti, T., Magnani, L., and Bardone, E. (2014). Camouflaging Truth: A Biological, Argumentative and epistemological Outlook from biological to Linguistic Camouflage. *Journal of Cognition and Culture,* 14: 67.

[113] Leonie, M. (2010). Resemblance and Camouflage in Graeco-Roman antiquity. *Sign Systems Studies. 38* (1-4): 180.

[114] Leonie, *ibid.,* 180.

[115] Leonie, *ibid.,* 180.

[116] Yang, R., Gao, D. X., Su, H., Wang, Z. W., Lei, Y., Fan, J., and Man, M. Y. (2014). To go hiding on camouflage meta-surfaces. *Applied Physics Letters,* 105 (21), 211602-1-211602-4.

[117] Webster, R. J. (2015). Does disruptive camouflage conceal edges and features? *Current Zoology,* 61 (4): 708.

[118] Bertolotti, et al., *ibid.,* 65.

[119] Bertolotti, et al., *ibid.,* 65.

[120] Webster, *ibid.,* 709.

[121] Webster, *ibid.,*708.

[122] Webster, *ibid.,* 708.

[123] Webster, *ibid.,* 708.

[124] Bertolotti et al., *ibid.,* 67.

[125] Forsyth, I. (2014). Designs on the desert: camouflage, deception and the militarisation of space. *Cultural Geographies,* 2 (2): 247.

[126] Leonie, *ibid.,* 182.

[127] Leonie, *ibid.,* 180.

[128] Wilke and Mata, *ibid.,* 533.

[129] Marshall, K. L. A., Philpot, K. E., Damas-Moreira, I., and Stevens, M. (2015). Intraspecific Colour Variation among Lizards in Distinct Island Environments Enhances Local camouflage. *PLOS ONE,* 10 (9): 1-20. DOI: 10.1371.: 13.

[130] Webster, *ibid.,* 709.

[131] Webster, *ibid.,* 709.

[132] Webster, *ibid.,* 708.

[133] Webster, *ibid.,* 104.

[134] Webster, *ibid.,* 711.

[135] Webster, *ibid.,* 711.

[136] Wilke and Rata, *ibid.,* 534.

[137] Webster, *ibid.,* 714.

[138] Forsyth, *ibid.,* 248.

[139] Troscianko, J., Lown, A. E., Hughes, A. E. and Stevens, M. (2013). Defeating Crypsis: detection and Learning of Camouflage Strategies. *PLOS ONE*, 8 (9):1.

[140] Troscianko, *ibid.,* 8.

[141] Leonie, *ibid.,* 168.

[142] Bertolotti et al., *ibid.,* 68.
[143] Bertolotti et al., *ibid.,* 68.
[144] Bertolotti et al., *ibid.,* 68.
[145] Millikan, R. G. (1984) *Language Thought and other Biological Categories*. Cambridge, MA: MIT Press.
[146] Bertolotti et al., *ibid.,* 69.
[147] Bertolotti et al., *ibid.,* 70.
[148] Bertolotti et al., *ibid.,* 72.
[149] Bertolotti et al., *ibid.,* 72.
[150] Bertolotti et al., *ibid.,* 72.
[151] Forsyth, *ibid.,* 248.
[152] Bertolotti et al., *ibid.,* 73
[153] Bertolotti et al., *ibid.,* 73.
[154] Bertolotti et al., *ibid.,* 74.
[155] Bertolotti et al., *ibid.,* 79.
[156] Bertolotti et al., *ibid.,* 77.
[157] Bertolotti et al., *ibid.,* 81.
[158] Bertolotti et al., *ibid.,* 82.
[159] Forsyth, *ibid.,* 261.
[160] Pinker, *ibid.,* 43.
[161] Baronchelli, Andrea, Nick Chater, Romualdo Pastor-Satorras, and Christiansen, M. H. (2012). The Biological Origin of Linguistic Diversity. *PLOS ONE,* 7 (10): 1
[162] Mossman, *ibid.,* 211.
[163] Harare, *ibid.,* 22.
[164] Fedurek, P., and Slocome, K. E. (2011). Primate Vocal Communication: A Useful Tool for Understanding Human Speech and Language Evolution? *Human Biology* 83 (2): 271.
[165] Jakobson, R. (1970). “La Linguistique.” In *Tendances principales de la recherché dans les sciences sociales et humanes.* Paris: UNESCO: 530.
[166] Kirby, S. (2007). The Evolution of Language. (Pp. 669–81) In *Oxford Handbook of Evolutionary Psychology*, edited by R. Dunbar, and L. Barrett,. Oxford: Oxford University Press.
[167] Smith et al., *ibid.,* 537.
[168] Fedurek and Slocome, *ibid.,* 153.
[169] Smith et al., *ibid.,* 538.
[170] Logan, R. (2004). *The Alphabet Effect: A Media Ecology Understanding of the Making of Western Civilization*. Cresskill, N. J.: Hampton Press.

[171] Goody, J. (1986). *The Logic of Writing and the Organisation of Society.* Cambridge: Cambridge University Press.
[172] Grosswiler, P. (2004). Dispelling the Alphabet Effect. *Canadian Journal of Communication* 29 (2): 5.
[173] Starte, L. (2006). *Echoes and Reflections: On Media Ecology as a Field of Study*. Cresskill, NJ: Hampton Press Communication Series: 45.
[174] Starte, *ibid.,* 46.
[175] Berwick, R and Chomsky, N. (2011). The biolinguistic program. The current state of its development. (Pp. 19-41). In *The Biolinguistic Enterprise: New Perspectives on the Evolution and Nature of the Human Language Faculty* edited by A-M, Di Sciullo, and C. Boeck. Oxford: Oxford University Press: 20-21.
[176] Dediu, D. 2011. Are Languages Really Independent From Genes? If Not, What Would a Genetic Bias Affecting Language Diversity Look Like? *Human Biology* 83 (2): 282.
[177] Schoenemann, T. P. (2009). Evolution of Brain and Language. *Language Learning* 59 (1): 175.
[178] Schoenemann, *ibid.,* 176.
[179] Schoenemann, *ibid.,* 163.
[180] Fedurek and Slocome, *ibid.,* 166.
[181] Leach, E. (1982). *Social Anthropology*. New York: Oxford University: 107.
[182] Kirby, *ibid.,* 2.
[183] Leach, *ibid.,* 107.
[184] Hauser, M. D., Chomsky, N., and Fitch, W. T. (2002). The Faculty of Language: What Is It, Who Has It, and How Did It Evolve? *Science* 298 (22): 1569–79.
[185] Kirby, *ibid.*
[186] Corballis, M. C. 2009. The Evolution of Language. *The Year in Cognitive Neuroscience: Annals of the New York Academy of Science* 1156: 19–43.
[187] Corballis, *ibid.,* 36.
[188] Lacoste-Badie, S and Droules, O. (2014) Advertising memory: the power of mirror neurons. *2014 NeuroPsychEconomics Conference Proceedings.* Ludwig Maximalian University, Munich, German. Retrieved from: Http://www.neuropsychoeconomics.org: 195.
[189] Richardson, P. J., and Boyd, R. 2010. Why Possibly Language Evolved. *Biolinguistics* 4 (2–3): 299.
[190] Richardson and Boyd, *ibid.,* 290.

[191] Johnson, N. (2012). *Simply Complexity: A Clear Guide To Complexity Theory*. London: One World Publications: 4.
[192] Yamauchi, H and Hashimoto, T. (2010). Relaxation of selection, niche construction, and the Baldwin effect of language evolution. *Artificial Life,* 16 (4): 271-287.
[193] Richardson, and Boyd, *ibid.,* 284.
[194] Richardson and Boyd, *ibid.,* 299.
[195] Smith, K., Brighton, H., and Kirby, S. 2003. Complex Systems in Language Evolution: The Cultural Emergence of Compositional Structure. *Advances in Complex Systems* 6 (4): 541.
[196] Blackmore, S. (1999). *The Meme Machine.* Oxford: Oxford University Press: 106
[197] Ekstig, *ibid.,* 184.
[198] MacWhinney, B. 2005. The Emergence of Linguistic Form in Time. *Connection Science* 17 (3–4): 191–211.
[199] Dediu, *ibid.,* 281.
[200] Fitch, W. T. (2010). The Evolution of Language. *New Scientist Instant Expert* I: viii. http://www.newscientist.com/data/doc/article/dn 19554/ instant_expert_6_-_the_evolution_of_language.pdf.: 203.
[201] Fitch, *ibid.,* 205.
[202] Ekstig, *ibid.,* 183.
[203] Ekstig, *ibid.,* 183.
[204] Chater, N,, and Christiansen, M. H.. (2009). Language Acquisition Meets Language Evolution. *Cognitive Science* 34: 1135-6.
[205] Chater and Christiansen, *ibid.,* 1141.
[206] Smith et al., *ibid.* 544.
[207] Raczaszek-Leonardi, J. (2010). Multiple Time-Scales of Language Dynamics: An Example from Psycholinguistics. *Ecological Psychology* 22: 269–85.
[208] Kuczynski, J-M. (2005). Can One Grasp Propositions Without Knowing a Language? *Teorema.* Vol. XXIV/2 : 44.
[209] Kuczynski, *ibid.,* 45
[210] Kuczynski, *ibid.,* 51.
[211] Kuczynski, *ibid,.* 52.
[212] Floyd, R. (2006). The Private Language Argument. *Philosophy Now*. 19.
[213] Floyd, *ibid.,* 19.
[214] Meyers, C. D. and Waller, S. (2009). Psychological investigations: the private language argument and inferences in contemporary cognitive science. *Synthese* 171: 135.

[215] Meyers and Waller, *ibid.,*155.
[216] Norman, R. (2000). Public Reasons and the 'Private Language' Argument. *Philosophical Investigations* 23: 4.: 29294.
[217] Norman, *ibid.,* 312.
[218] Schoenemann, *ibid.,* 163.
[219] Fedurek and Slocome, *ibid.,* 166.
[220] Leach, *ibid.,* 107.
[221] Seamon, J. G., Brody, N., and Kauff, D. M. (1983). Affective discrimination of stimuli that are not recognised: II. Effect of delay between study and test. *Bulletin of the Psychonomic Society*, 21.: 188.
[222] Bostrom, N. (2003). Are you living in a computer simulation? *Philosophical Quarterly*, 53: 211: 243-255.
[223] Neppe, V. M. (1983). *The Psychology of Déjà vu: Have I been here before?* Johannesburg, South Africa: Witwaters and University Press.: 3.
[224] Brown, A. S. (2003). A Review of the Déjà vu Experience. *Psychological Bulletin.* 129 (3): 397.
[225] Brown, *ibid.,* 409.
[226] Ramachandran, V. S. (2011). *The Tell-Tale Brain.* London: Windmill Books: 284.
[227] Ramachandran, *ibid.,* 284.
[228] Moulin, C. J. A and Chauvel, P. (2010). Déjà vu: Insights from the dreamy state and memory research. (Pp. 206-235). In D. B. Wright and G. Davies (Eds). New frontiers in applied memory Howe, UK: Psychology Press: 2010, p. 214.
[229] Moulin and Chauvel, *ibid.,* 214.
[230] Brown, *ibid.,* 395.
[231] O'Connor, A. R., and Moulin, C. J. A. (2010). Recognition Without Identification, Erroneous Familiarity, and Déjà vu. *Current Psychiatric Reports*. 12.
[232] Cleary, A. M. (2008). 'Recognition Memory, Familiarity, and Déjà vu Experiences. *Current directions in Psychological Science.* 17 (5): 353.
[233] Cleary, *ibid.,* 354.
[234] Brown, *ibid.,* 394.
[235] Kohr, R. L. (1980). A survey of psi experiences among members of a special population. *Journal of the American Society for Psychical Research*, 74: 246-249.
[236] Brown, *ibid.,* 398.
[237] Neppe, *ibid.*
[238] Neppe, *ibid.*

[239] Brauer, R., Harrow, M., and Tucker, G. J. (1970). Depersonalization phenomena in psychiatric patients. *British Journal of Psychiatry*, 117: 509-515.

[240] Chapman, A. H., and Mensch, I. N. (1951). Déjà vu experience and conscious fantasy in adults. *Psychiatric Quarterly Supplement*, 25: 168-169.

[241] Richardson, T. F., and Winokur, G. (1967). Déjà vu in psychiatric and neurosurgical patients. *Archives of General Psychiatry*, 17: 622-625.

[242] Brown, *ibid.,* 401.

[243] Chapman and Mensch, *ibid.*

[244] Titchener, E. E. (1924). *A Beginner's Psychology.* New York: Macmillan: 187.

[245] Cleary, A. M., Ryals, A. J., and Nomi, J. S. (2009). Can déjà vu result from similarity to a prior experience? Support for the similarity hypothesis of déjà vu. *Psychonomic Bulletin and Review*. 16 (6): 1082.

[246] Moulin and Chauvel, *ibid.*

[247] Neppe, *ibid.*

[248] Brown, *ibid.* 394.

[249] Brown, *ibid.,* 398.

[250] Brown, *ibid.,* 402.

[251] Brown, *ibid.,* 403

[252] De Nayer, A. (1979). Déjà vu: Elaboration of a hypothetical model of explanation. *Psychiatria Clinica*, 12: 92-96.

[253] Pashler, H. (1994). Dual-task interference in simple tasks: Data and theory. *Psychological Bulletin*, 116: 220-244.

[254] Hughlings-Jackson, J. (1988). On a particular variety of epilepsy ("intellectual aura"), one case with symptoms of organic brain disease. *Brain*, 11: 179-207.

[255] Halgren, E., Walter, R. D., Cherlow, D. G., and Crandall, P. H. (1978). Mental phenomena evoked by electrical stimulation of the hippocampal formation and amydala. *Brain*, 101: 83-117.

[256] Bancaud, J., Brunet-Bourgin, F., Chauvel, P., and Halgren, E. (1994). Anatomical origin of déjà vu and vivide memories in human temporal lobe epilepsy. *Brain*, 117: 71-90.

[257] Wild, E. (2005). 'Déjà vu in neurology.' *Journal of Neurology*, 252:1: 1-7.

[258] Grasset, J. (1904). La sensation du 'déjà vu.' *Journal de Psychologie, Normale et Pathologique*, 1:17.

[259] Ephron, R. (1963). Temporal perception, aphasia, and déjà vu. *Brain*, 86: 403-424.

[260] Humphrey, G. (1923). *The story of man's mind.* Boston: Small, Maynard.

[261] Fleminger, S. (1991). The déjà vu experience: Remembrance of things past?: Comment. *American Journal of Psychiatry*, 148. 1418.

[262] Linn, L. (1953). Psychological implications of the 'activating system.' *American Journal of Psychiatry*, 110: 61-65.

[263] Brown, *ibid.*, 407.

[264] Mack, A.., and Rock, I. (1998). *Inattentional blindness*. Cambridge, MA: MIT Press.

[265] Brown, *ibid.*, 407.

[266] Dixon, N. F. (1971). *Subliminal perception: The nature of a controversy.* London: McGraw-Hill.: 106.

[267] O'Connor, A. R., and Moulin, C. J. A. (2010). Recognition Without Identification, Erroneous Familiarity, and Déjà vu. *Current Psychiatric Reports*. 12: 165.

[268] Adewumi, A. Kagamba, J, and Alochukwu, A. (2015). Application of Chaos Theory in the Prediction of Motorised Traffic Flows on Urban Networks. Mathematical Problems in Engineering. Http://dx.doi.org/10.1155/2016/5656734:_5.

[269] Brachthauser, C. (2011). Explaining global governance – a complexity perspective. *Cambridge Review of International Affairs.* 24 (2): 222.

[270] Lorenz, E. N. (1972). Predictability: Does the Flap of a Butterfly's wings in Brazil Set Off a Tornado in Texas? American Association For The Advancement of Science, 139th Meeting. Retrieved from: http://eaps4.mit.edu/research/Lorenz/Butterfly_1972.pdf, 234.

[271] Gilpin, D. R., and Murphy, P. J. (2008). *Crisis management in a Complex World.* New York, NY: Oxford University Press.

[272] Mossman, *ibid,* 71.

[273] Gilpin and Murphy, *ibid.*

[274] Crutchfield, J. P. and Machta, J. (2011). Introduction to focus issue on 'randomness, structure, and causality: measures of complexity from theory to applications.' *Chaos.* 21 (3): 037101. doi: 10.1063/1.3643065.

[275] Mossman, *ibid.*, 72.

[276] Ekstig, *ibid.*, 177-178.

[277] Gould, S. J. (1996). *Full House: The Spread of Excellence From Plato to Darwin.* New York: Harmony House: 173.

[278] Ekstig, *ibid.*, 179.

[279] Ekstig, *ibid.,* 180.

[280] Davies, P. C. W. (2013). Directionary principles from cancer to cosmology. In C. H. Lineweaver, Davies P., and Ruse, M. (Eds.) *Complexity and the arrow of time.* Cambridge: Cambridge University Press.

[281] Ekstig, *ibid.,* 182.

[282] Adewumi, et al., *ibid.,* 6.

[283] Adewumi, et al., *ibid.,* 6.

[284] Adewumi, et al., *ibid.,* 6.

[285] Brachthauser, *ibid.,* 223.

[286] Johnson, *ibid.,* 3-4.

[287] Bryne, D. S. (1998). *Complexity Theory and the Social Sciences.* London: Routledge: 5.

[288] Gilpin, D. R. and Murphy, P. (2008). *Crisis Management in a Complex World.* New York: Oxford University Press: 26.

[289] Brachthauser, *ibid.,* 34.

[290] Briggs, J. and Peat, F. D. (1989) *Turbulent Mirror. An Illustrated Guide to Chaos Theory and the Science of Wholeness.* New York: Harper and Row: 154.

[291] Sanger, M. and Giddings, M. M. (2012). Teaching Note A Simple Approach To Complexity Theory. *Journal of Social Work Education*, 48 (2): 371.

[292] Alemi, M., Daftarifard, P., and Patrut, B. (2011). The Implication of Chaos/Complexity Theory into Second Language Acquisition. *BRAIN. Broad Research in Artificial Intelligence and Neuroscience*. 2 (2): 37.

[293] Alemi, et al., *ibid.,* 25.

[294] Bousquet, A. and Curtis, S. (2011). Beyond Models and metaphors: complexity theory, systems thinking and international relations. *Cambridge Review of International Affairs.* 24 (1). 45.

[295] Bousqet and Curtis, *ibid.,* 46.

[296] Valle, V. (2000). Chaos, complexity, and deterrence. Retrieved from https://www.au.af.mil/au/awc/awcgate/ndu/valle.pdf.

[297] Brachthauser, *ibid.,* 227.

[298] Ekstig, *ibid.,* 175.

[299] Mossman, *ibid.,* 73.

[300] Ekstig, *ibid.,* 175.

[301] Sanger and Giddings, *ibid.,* 372.

[302] Diaz-Parra, O., and Ruiz-Vanoye, J. A. (2011). An Overview of the Theory of Instances of Computational Complexity. *International*

Journal of Combinatorial Optimization Problems and Informatics. 2 (2): 21.

[303] Diaz-Parra and Ruiz-Vanoye, *ibid.,* 22.

[304] Diaz-Parra and Ruiz-Vanoye, *ibid.,* 22.

[305] Diaz-Parra and Ruiz-Vanoye, *ibid.,* 22.

[306] Proponen, P., Ko, K-I, Schoning, U., and Watanabe, O. (1994). *Instance Complexity. Journal of the ACM* 41(1):96-121

[307] Aldrich, R. M. (2008). From Complexity Theory to Transactionalism: Moving Occupational Science Forward in Theorising the Complexities of Behaviour. *Journal of Occupational Science*, 15 (3):148.

[308] Aldrich, *ibid.,* 148.

[309] Aldrich, *ibid.,* 148.

[310] Brachthauser, *ibid.,* 224.

[311] Brachthauser, *ibid.,* 224.

[312] Brachthauser, *ibid.,* 227.

[313] Bryne, *ibid.,* 39.

[314] Bryne, *ibid.,* 53.

[315] Mossman, *ibid.,* 73.

[316] Brachthauser, *ibid.,* 225.

[317] Sanger and Giddings, *ibid.,* 370.

[318] Sanger and Giddings, *ibid.,* 371.

[319] Brachthauser, *ibid.,* 223.

[320] Mason, M. (2008). Complexity Theory and the Philosophy of Education. *Education Philosophy and Theory*. 40 (1): 5.

[321] Mason, *ibid.,* 5.

[322] Mason, *ibid.,* 10.

[323] Sanger and Giddings, *ibid.,* 373.

[324] Mason, *ibid.,* 12.

[325] Mason, *ibid.,* 12.

[326] Alemi, et al., *ibid.,* 39.

[327] Woesler, R. (2007). Complexity Theory Gives Hints for Final Answers to Fundamental Physical Questions. *Physics Essays.* 20 (4): 533.

[328] Woesler, *ibid.,* 528.

[329] Harare, *ibid.* 9.

[330] Stevenson, I. H, Cronin, B., Sur, M, and Kording, K. P. (2010) Sensory Adaptation and Short Term Plasticity as Bayesian Correction for a Changing Brain. *PLOS ONE* 5(8): e12436.

[331] Ramachandran, *ibid.,* 270.

[332] Trenholm, S., and Jensen, A. (2004). *Interpersonal communication*. New York: Oxford University Press: 163.
[333] Trenholm and Jensen, *ibid.,* 164-165.
[334] Wilke and Mata, *ibid.,* 532.
[335] Eunson, B. (2006). *Communicating in the 21st Century.* Australia: John Wiley & Sons Australia: 433-435.
[336] Van Doorn, E. A., Heerdink, M. W., and Van Kleef, G. A. (2012). Emotion and the construal of social situations: Inferences of cooperation versus competition from expressions of anger, happiness and disappointment. *Cognition and Emotion,* 26 (3): 444.
[337] Van Doorn et al., *ibid.,* 444.
[338] Van Doorn et al., *ibid.,* 454.
[339] Han, T. A. (2013). *Intention Recognition, Commitment and Their Roles in the Evolution of Cooperation. From Artificial Intelligence Techniques to Evolutionary Game Theory Models.* London: Springer: 6.
[340] Rindfleisch, A. and Moorman, C. (2003). Interfirm co-operation and customer orientation,' *Journal of Marketing Research*, 40 (4): 421-36.
[341] Luo, X., Slotegraaf, R. J., and Pan, X. (2006). Cross-Functional 'Cooperation': The Simultaneous Role of Cooperation and Competition Within Firms, *Journal of Marketing*, 70 (2): 69.
[342] Luo et al., *ibid.,*70.
[343] Franken, R. E. and Brown, D. J. (1996). The need to win is not adaptive: The need to win, coping strategies, hope and self-esteem. *Personality and Individual Differences.* 20 (6): 805.
[344] Franken and Brown, *ibid.,* 807.
[345] Franken and Brown, *ibid.,* 805.
[346] Franken and Brown, *ibid.*
[347] Zhang, H., Shu, C., Jiang, X., and Malter, A. J. (2010). Managing Knowledge for innovation: The Role of Cooperation, Competition, and Alliance Nationality. *Journal of International Marketing.* 18 (4): 78.
[348] Zhang et al., *ibid,* 86.
[349] Peng, T-J A., Pike, S., Yang, J C-H, and Roos, G. (2012). Is Cooperation with Competitors a Good Idea? An Example in Practice. *British Journal of Management,* 23 (4): 532.
[350] Peng, et al., *ibid.,* 533.
[351] Bengtsson, M and Kock, S. (2000). "Coopetition" in business networks – To cooperate and compete simultaneously. *Industrial Marketing Management* 26 (5): 411-461.
[352] Peng et al., *ibid.,* 535.

[353] Peng, et al., *ibid.,* 536.
[354] Peng, et al., *ibid.,* 548.
[355] Peng, et al., *ibid.,* 584.
[356] Rubin, P. H. (2014). Emporiophobia (Fear of Markets): Cooperation or Competition? *Southern Economic Journal,* 80 (4): 876.
[357] Rubin, *ibid.,* 878.
[358] Rubin, *ibid.,* 880.
[359] Rubin, *ibid.,* 881.
[360] Rubin, *ibid.,* 883.
[361] Cited in Berg, *ibid.,* 176.
[362] Berg, *ibid.,* 180.
[363] Roosevelt, F. D. (March 3, 1912). Speech at the People's Forum in Troy, New York.
[364] Malhotra, D. (2009). The Desire to Win: The Effects of Competitive Arousal on Motivation and Behaviour. *Organizational Behaviour and Human Decision Processes,* 111 (2): 139.
[365] Ku et al., *ibid.*
[366] Messick and Thorngate, *ibid.*
[367] Malhotra, *ibid.,* 140.
[368] Malhotra, *ibid.,* 140.
[369] Jorde, T M., and Teece D. J. (1989). Innovation, Cooperation, and Antitrust: Balancing Competition and Cooperation, *High Technology Law Journal,* 4 (1): 26.
[370] Jorde and Teece, *ibid.,* 26.
[371] Jorde and Teece, *ibid.,* 26.
[372] Jorde and Teece, *ibid.,* 26.
[373] Han, *ibid.,* 60.
[374] Jorde and Teece, *ibid.,* 26.
[375] Ims and Jakobsen, *ibid.,* 19.
[376] Ims and Jakobsen, *ibid.,* 19.
[377] Ims and Jakobsen, *ibid.,* 19.
[378] Ima and Jakobsen, *ibid.,* 19.
[379] Ims and Jakobsen, *ibid.,* 21.
[380] Ims and Jakobsen, *ibid.,* 21.
[381] Capra, F. (1982) *The Turning Point: Science, Society, and the Rising Culture*, Simon and Schuster.
[382] Ims and Jokobsen, *ibid.,* 22.

[383] Donne, J. (1624). *Devotions upon emergent occasions and several steps in my sickness - Meditation XVII.* Retrieved from: http://www.online-literature.com/donne/409/.

[384] Han, *ibid.,* 39.

[385] Tauer and Harackiewicz, *ibid.,* 226.

[386] Shaomeng, *ibid.,* 274-275.

[387] Shaomeng, *ibid.,* 275.

[388] Shaomeng, *ibid.,* 276.

[389] Shaomeng, *ibid.,* 279.

[390] Shaomeng, *ibid.,* 281.

[391] Shaomeng, *ibid.,* 281.

[392] Osarenkhoe, A. (2010). A study of inter-firm dynamics between competition and cooperation – A co-opetition strategy. *Database Marketing and Customer Strategy Management.* 17 (3/4): 201.

[393] Osarenkhow, *ibid.,* 201.

[394] Osaenkhow, *ibid.,* 204.

[395] Osarenkhow, *ibid.,* 204.

[396] Cangello cited in Osarenkhoe, *ibid.,* 205.

[397] Osarenkhow, *ibid.,* 205.

[398] Osatrenkhow, *ibid.,* 217.

[399] Osarenkhoe, *ibid.,* 217.

[400] Osarenkhow, *ibid.,* 217.

[401] Brittain, S. (2012). Interviewing Skills: Building a Solid Structure. *People Management:* 30.

[402] Morgan, C. A., Rabinowitz, Y., Leidy, R., and Coric, V. (2014). Efficacy of Combining Interview Techniques in Detecting Deception Related to Bio-Threat Issues. *Behavioural Sciences and the Law,* 32: 269.

[403] Morgan et al., *ibid.,* 270.

[404] Morgan et al., *ibid.,* 284.

[405] Windle, C., McFann, H., and Ward, J. (1955). The Effect of Various Interview Techniques in Evoling Fear Responses. *Journal of Clinical Psychology.* 11 (2): 171.

[406] Brown, C., Lloyd-Jones, T. J. and Robinson, M. (2008). Eliciting person descriptions from eyewitnesses: A survey of police perceptions of eyewitness performance and reported use of interview techniques. *European Journal of Cognitive Psychology.* 20:3, 556.

[407] Dando, C. J., Wilcock, R., Behnkle, C., and Milne, R. (2011). Modifying the cognitive interview: countenancing forensic application by enhancing practicability. *Psychology, Crime and Law,* 17:6:491-511.

[408] Memon, A., Wark, L., Bull, R., and Koehnken, G. (1997). Isolating the effects of the cognitive intyerview techniques. *British Journal of Psychology,* 88: 182.

[409] Memon et al., *ibid.,* 180.

[410] Davis, M. R., McMahon, M., and Greenwood, K. M., 2004. The Role of Visual Imagery in the Enhanced Cognitive Interview: Guided Questioning Techniques and Individual Differences. *Journal of Investigative Psychology and Offender Profiling,* 1: 48.

[411] Davis et al., *ibid.,* 48.

[412] Davis, et al., *ibid.,* 35.

[413] Dando, et al., *ibid.,* 126.

[414] Wysman, L., Scoboria, A., Gawrylowicz, J., and Memon, A. (2014). The Cognitive Interview Buffers the Effects of Subsequent Repeated Questioning in the Absence of Negative Feedback. *Behavioural Sciences and the Law,* 32 (2): 207.

[415] Wysman, et al., *ibid.,* 208.

[416] Wysman et al., *ibid.,* 208.

[417] Gudjonsson, G. H. (2003). Psychology Brings Justice: The Science of Forensic Psychology. *Criminal Behavior and Mental Health,* 13: 159-167.

[418] Wyseman et al., *ibid.,* 209.

[419] Wyseman et al., *ibid.,* 217.

[420] Kniazeva, M. (2005). How Many Wings Does a Butterfly Have? Or Stimuli As An Elicitation Technique For In-Depth Interviews.*Advances in Consumer Research,* 32: 540-544.

[421] Memon, et al., *ibid.,* 182.

[422] Bekerian, D. A. and Dennett, J. L. (1993). The Cognitive Interview Technique: Reviving the Issues. *Applied Cognitive Psychology.* 7: 276.

[423] Bekerian and Dennett, *ibid.,* 276.

[424] Nigro, G., and Neisse, V. (1983). Point of view of Personal memories. *Cognitive Psychology,* 15: 467-482.

[425] Memon et al., *ibid.*

[426] Gesielman, R. E., Fisher, R. P., MacKinnon, D. P., and Holland, H. L. (1985). Eyewitness memory enhancement in the cognitive interview: Cognitive retrieval mnemonics versus hypnosis. *Journal of Applied Psychology,* 70., 288.

[427] Gesielman et al., *ibid.,* 288.
[428] Gesielman et al., *ibid.,* 288.
[429] Windle, et al., *ibid.,* 173
[430] Smith.L. L. and Bull, R. (2014). Exploring the Disclosure of Foresnic Evidence in Police Interviews with Suspects. *Journal of Police and Criminal Psychology,* 29: 85.
[431] May, L., Granhag, P. A., and Oleszkiewicz, S. (2014). Eliciting Intelligence Using the Scharff-Technique: closing in on the Confirmation/Disconfirmation-Tactic. *Journal of Investigative Psychology and Offender Profiling,* 11: 137-138.
[432] Chaudoir, S. R. and Fisher, J. D. (2010). The Disclosure Processes Model: Understanding Disclosure Decision Making and Post Disclosure Outcomes Among People Living With a Concealable Stigmatized Identity. *Psychological Bulletin*, 136 (2): 236.
[433] Chaudoir and Fisher, *ibid.,* 276.
[434] Ziv-Beiman, S. (2013). Therapist Self-Disclosure as an Integrative Intervention. *Journal of Pyschotherapy Integration*, 23 (1): 60.
[435] Ziv-Beiman, *ibid.,* 61-62.
[436] Ziv-Beiman, *ibid.,* 63.
[437] Ziv-Beiman, *ibid.,* 63.
[438] Littlejohn, S. W. (1983). *Theories of Human Communication* (2nd Ed.) Belmont, California: Wadsworth.: 195.
[439] Garrison, A. M., Kahn, J. H., Sauer, E. M., and Florczak, M. A. (2012). Disentangling the Effects of Depression Symptoms and Adult Attachment on Emotional Disclosure. *Journal of Counselling Psychology*, 59 (2): 230.
[440] Garrison, et al., *ibid.,* 230.
[441] Garrison, et al., *ibid.,* 230.
[442] Garrison, et al., *ibid.,* 231.
[443] Garrison, et al., *ibid.,* 231.
[444] Garrison, et al., *ibid.,* 231.
[445] Garrison, et al., *ibid.,* 231.
[446] Foynes, M. M. and Freyd, J. J. (2013). An Exploratory Evaluating Responses to the Disclosure of Stressful Life Experiences as They Occurred in Real Time. *Psychological Trauma: Theory, Research, Practice, and Policy*, 5 (3): 295.
[447] Barnett, J. E. (2011). Psychotherapist Self-Disclosure: Ethical and Clinical Considerations. *Psychotherapy*, 48 (4): 315.

[448] Chaudoir, S. R., and Quinn, D. M. (2010). Revealing Concealable Stigmatized Identities: The Impact of Disclosure Motivations and Positive First-Disclosure Experiences on Fear Disclosure and Well-Being. *Journal of Social Issues,* 66 (3): 570-584.

[449] Chaudoir and Quinn, *ibid.,* 571.

[450] Chaudoir and Quinn, *ibid.,* 572.

[451] Chaudoir and Quinn, *ibid.,* 572.

[452] Chaudoir and Quinn, *ibid.,* 573..

[453] Sah, S., Cain, D. M., and Loewenstein, G. (2013). The Burden of Disclosure: Increased Compliance With Distrusted Advice. *Journal of Personality and Social Psychology,* 104 (2): 289.

[454] Sah, et al., *ibid.,* 289.

[455] Sah, et al., *ibid.* 289.

[456] Sah, et al., *ibid.* 290.

[457] Sah, et al., *ibid.* 290.

[458] Sah, et al., *ibid.,* 290.

[459] Vitak, J. (2012). The Impact of Context Collapse and Privacy on Social Network Site Disclosures. *Journal of Broadcasting and Electronic Media,* 56 (4): 452.

[460] Boyd (2008) cited in Vitak, *ibid.,* 453.

[461] Nosko, A., Wood, E., and Molema, S. (2010). All about me: Disclosure in online social networking profiles: The Case of FACEBOOK. *Computers in Human Behaviour,* 26: 406-418.

[462] Newness, K., Steinart, J., and Viswesvaran, C. (2012). Effects of Personality on Social Network Disclosure: Do Emotionally Intelligent Individuals Post Inappropriate Content? *Psychological Topics*, 21 (3): 473.

[463] Littlejohn, *ibid.,* 196.

[464] Jourard, S. (1968). *Disclosing Man to Himself.* New York: Van Nostrand: 154.

[465] Goleman, D. (1995). *Emotional Intelligence: Why It Can Matter More Than IQ.* New York: Bantam Books.

[466] Feinberg, M., Willer, R., and Keltner, D. (2012). Flustered and faithful: Embarrassment as a signal of prosociality. *Journal of Personality and Social Psychology*, *102*(1). 81–97: 81.

[467] Keltner, D. and Buswell, B. N. (1997). Embarrassment: Its distinct form and appeasements functions. *Psychological Bulletin*, *122*(3): 250–270.

[468] Hawk, S. T., Fischer, A. H., and Van Kleef, G. A. (2011). Taking your place or matching your face: Two paths to empathic embarrassment. *Emotion, 11*(3), 153-161.
[469] Hawk, et al., *ibid.*
[470] Hawk et al., *ibid.*
[471] Feinberg, et al., *ibid.,* 81.
[472] Feinberg, et al., *ibid.*
[473] Darby, R. S. and Harris, C. R. (2010). Embarrassment's effect on facial processing.*Cognition and Emotion, 24*(7): 1250-1258.
[474] Feinberg et al., *ibid..*
[475] Keltner, and Buswell, *ibid.,* 250–270.
[476] Keltner and Buswell, *ibid.*
[477] Feinberg et al., *ibid.*
[478] Feinberg et al., *ibid.*
[479] Darby and Harris, *ibid.,* 1250-1258
[480] Keltner and Buswell, *ibid.*
[481] Hawk, et al. *ibid.*
[482] Hawk et al., *ibid.*
[483] Hawk et al., *ibid.*
[484] Feinberg et al., *ibid.,* 81.
[485] Keltner and Buswell, *ibid.*
[486] Robbins, B. D. and Parlavecchio, H. (2006). The unwanted exposure of the self: A phenomenological study of embarrassment. *The Humanistic Psychologist*, *34*(2): 321–345.
[487] Robbins and Parlavecchio, *ibid..*
[488] Frijda, N. H. (1988). The Laws of Emotion. *American Psychologist, 43*(5): 349.
[489] Gao, L., Wheeler, C. S., and Shiv, B. (2008). The "Shaken Self": Product choices as a means of restoring self-view confidence. *Journal of Consumer Research, 36* (1): 29–38.
[490] Zhong, C-B., and Liljenquist, K. (2006), Washing away your sins: Threatened morality and physical cleansing. *Science, 313:* 1451–52.
[491] Gao, et al., *ibid.*
[492] Gao et al., *ibid.*
[493] Gao et al., *ibid.*
[494] Gao et al., *ibid.*
[495] Wan, E. W., and Rucker, D. D. (2012). Confidence and construal framing: When confidence increases versus decreases information processing. *Journal of Consumer Research, 39* (4): 977–992.

[496] Wan and Rucker, *ibid.*

[497] Wan and Rucker, *ibid.*

[498] Wan and Rucker, *ibid.*

[499] Briñol, P., Petty, R. E., Valle, C., Rucker, D. D., and Becerra, A. (2007). The effects of message recipients' power before and after persuasion: A self-validation analysis. *Journal of Personality and Social Psychology, 93*(6): 1040–53.

[500] Wan and Rucker, *ibid.,* 978.

[501] Wan and Rucker, *ibid.,* 978.

[502] Wan and Rucker, *ibid.,* 978.

[503] Wan and Rucker, *ibid.,* 978.

[504] Van Zant, A. B., and Moore, D. A. (2013). Avoiding the pitfalls of overconfidence whilebenefiting from the advantages of confidence. *California Management Review, 55*(2): 5–23.

[505] Van Zant and Moore, *ibid.*

[506] Fisher, C. D., Minbashian, A., Beckmann, N. and Wood, R. E. (2013). Task appraisals, emotions, and performance goal orientation. *Journal of Applied Psychology* 98 (2): 364-373.

[507] Fisher et al., *ibid.*

[508] Ratcliff, R., and Starns, J. J. (2013). Modelling confidence judgments, response times, and multiple choices in decision making: Recognition memory and motiondiscrimination. *Psychological Review, 120*(3): 697–719.

[509] Koriat, A. (2011). Subjective confidence in perceptual judgments: A test of the self-consistency model. *Journal of Experimental Psychology, 140*(1): 117–139.

[510] Koriat, *ibid.*

[511] Koriat, *ibid.*

[512] Baranski, J. V. and Petrusic, W. M. (1994). 'judgments.' Perception *and Psychophysics*, 55(4): 412-28.

[513] Koriat, *ibid.*

[514] Koriat, *ibid.*

[515] Selmeczy, D., and Dobbins, I. G. (2013). Relating the content and confidence of recognition judgments. *Journal of Experimental Psychology: Learning, Memory and cognition, 40*(1): 66-85.

[516] Jones, E. E., Williams, K. D., and Brewer, N. (2008). "I had a confidence Epiphany!" Obstacles to combatting post identification confidence. *Law and Human Behavior, 32*: 164–176.

[517] Metcalfe, J. and Finn, B. (2011). People's hypercorrection of high-confidence errors: Did they know it all along? *Journal of Experimental Psychology: Learning, Memory and Cognition*, *37*(2): 437–48.

[518] Metcalfe and Finn, *ibid.*

[519] Metcalfe and Finn, *ibid.*

[520] Buratti, S., and Allwood, C. M. (2012). Improved realism of confidence for an episodic memory event. *Judgment and Decision Making, 7*(5): 590-601.

[521] Buratti and Allwood, *ibid.*

[522] Buratti and Allwood, *ibid..*

[523] Nielsen, T. A. and Stenstrom, P. (2005). What are the memory sources of dreaming? *Nature, 437*: 1286-1289.

[524] Fosse, M. J., Fosse, R., Hobson, J. H., and Stickgold, R. J. (2003) Dreaming and episodic memory: A functional dissociation? *Journal of Cognitive Neuroscience*, *15*: 1-9.

[525] Lee, R. L. M.. (2010) Forgotten fantasies? Modernity, Re-enchantment, and Dream Consciousness. *Dreaming*, *20*: 288-304.

[526] *LaBerge, S. (1990). Lucid Dreaming: Psychophysiological Studies of Consciousness During REM Sleep.* (Pp. 109-126). In Bootzen, R.R., Kihlstrom, J.F. and Schacter, D.L., (Eds.) *Sleep and Cognition.* Washington, D.C.: American Psychological Association.

[527] Hanfling, O. (1984). What Does The Private Language Argument Prove? *The Philosophical Quarterly* 34 (137): 468-481.

[528] Doyon, J., Carrier, J., Simard, A., Tahar, A. H., Morin, A., Benali, H., and Ungerleider, L. G. (2005). Motor memory: Consolidation-based enhancement effect revisited. *Behavioural and Brain Sciences*, *28*: 68.

[529] Jenkins, J. G., and Dallenbach, K. M. (1924). Obliviscence during sleep and waking. *American Journal of Psychology. 35:* 605–612.

[530] Moruzzi, G., and Magoun, H. W. (1949). 'Brain stem reticular formation and activation of the EEG.' *Electroencephalogr. Clin. Neuro. 1*: 455-73.

[531] Greenberg, R. and Pearlman, C. (1974). Cutting the REM nerve: An approach to the adaptive role of REM sleep. *Perspectives in Biology and Medicine*, 17: 513-521.

[532] Siegel, A. B. (2005). The incredible shrinking sleep-learning connection. *Behavioural and Brain Science,* 28: 82-83.

[533] Wamsley, E. J. and Stickgold, R. (2010). Dreaming and offline memory processing,' *Current Biology*, *20*, R1010-R1013.

[534] *Flanagan,* O. (*1995*) Neuroscience and Dreams: The *Spandrels of Sleep*, *Journal of Philosophy,* XCII, 1:5-27.

[535] Hahn, T. T. G., McFarland, M. J., Berberich, S., Sakmann, B., and Mehta, M. R. (2012). Spontaneous persistent activity in entorhinal cortex modulates cortico-hippocampal interaction *in vivo. Nature Neuroscience, 15,* 1531–1538. doi:10.1038/nn.3236.

[536] Markowitsch, H. J. and Staniloiu, A. (2013). Commentary: The spaces left over between REM sleep, dreaming, hippocampal formation, and episodic autobiographical memory. *Behavioural and Brain Sciences*, *36*, 34-35.

[537] Groeger, J. A., and Dijk, D-J. (2005). Consolidating consolidation? Sleep stages, memory systems, and procedures, *Behavioural and Brain Sciences*, *28*, 73.

[538] Hobson, J.M., McCarley, R. W., and Wyzinski, P. W. (1975). Sleep cycle oscillation: Reciprocal discharge by two brainstem neuronal networks, *Science*, *189*: 55-58.

[539] Lee, *ibid.,* 288.

[540] Lee, *ibid.,* 288.

[541] Lee, *ibid.,* 293.

[542] Hobson, et al., *ibid.*

[543] Wamsley, and Stickgold, *ibid.,* R10101.

[544] Nielsen and Stenstrom, *ibid.,* 1286.

[545] Lee, *ibid.*

[546] Nielsen and Stenstrom, *ibid.,* 1287.

[547] Nielsen and Stenstrom, *ibid.,* 1287.

[548] Nielsen and Stenstrom, *ibid.,* 1287.

[549] Nielsen and Stenstrom, *ibid.,* 1288.

[550] Nielsen and Stenstrom, *ibid.,* 1288.

[551] Wamsley, and Stickgold, *ibid.,* R10102.

[552] Fosse, et al., *ibid.*, 6.

[553] Barnett, *ibid.,* 9.

[554] Wamsley and Stickgold, *ibid.,* R1013.

[555] Wamsley, E. J., and Antrobus, J. S. (2009). The expression of trace conditioning during non-REM sleep and its relation to subjective experience. *Neurobiology of Learning and Memory*. *92*, 283.

[556] Wamsley and Antrobus, *ibid.,* 289.

[557] Walker, M. P. (2005). A refined model of sleep and the time course of memory formation, *Behavioural and Brain Sciences*, *28*, 64.

[558] Tononi, G. and Cirelli, C. (2005). Sleep and synaptic homeostatisis. *Behavioural Brain Sciences*, *28*, 85.

[559] Maestes, K.L. and Rude, S. S. (2012). The Benefits of Expressive Writing on autobiographical Memory Specificity: A randomised controlled trial, *Cognitive Therapy and Research* 36: 235.

[560] Maestes and Rude, *ibid.,* 244.

[561] Schacter, D. L. (2001). *The Seven Sins of Memory: How the mind forgets and remembers*, Boston: Houghton Mifflin Company: 4.

[562] Loughran, E. (2010). Defective Memories, Deception, and the Writing Process: Montaigne's attempt at truth in essay, I (9) *Neophilogus* 94: 33–41.

[563] Pigni, A. (2006). Symbolic communication and silence in an informal interpersonal helping process (extract from the film Chocolat by Lass Hallström). *Psychodynamic Practice,* 12 (4).

[564] Kennedy, E. and Charles, S. C. (1990). *On Becoming a Counsellor,* Second Edition, Dublin: Gill and Macmillan: 66-67.

[565] Sands, S. H. (2010). On the Royal Road Together: The Analytic Function of Dreams in Activating Dissociative. *Unconscious Communication,* 20 (4): 358.

[566] Modell, A. H. (1997). The synergy of memory, affects and metaphor. *Journal of Analytic Psychology,* 42 (1): 105-117.

[567] Jacobs, T. J. (2001). On Unconscious Communications and Covert Enactments: Some Reflections on Their Role in the Analytic Situation. *Psychoanalytic Inquiry.*, 21 (1): 5.

[568] Jacobs, *ibid.,* 8.

[569] Dimberg, U., Thunberg, M., and Elmehed, K. (2000). Unconscious Facial Reactions to Emotional; Facial Expressions. *Psychological Science.* 11 (1): 86.

[570] Deecke, L. (2012). There Are conscious and unconscious Agendas in the Brain and Both Are Important – Our Will Can Be Conscious as well as unconscious. *Brain Science,* 2: 405.

[571] Deecke, *ibid.,* 406.

[572] Zajonc, R. B. (1980). Feeling and thinking: Preferences need no inferences. *American Psychologist*, 35: 151-175.

[573] Zajonc, *ibid.*

[574] Knoblauch, S. H. (2005). Body Rhythms and the unconscious Toward an Expanding of Clinical Attention. *Psychoanalytic Dialogues* 15 (6): 807-827.

[575] Lacoste-Badie and Droules *ibid.,* 195.

[576] Deecke, *ibid.,* 413.

[577] Deecke, *ibid.,* 413.

[578] Waugaman, R. M. (2007). Unconscious Communication in Shakespeare: "Et Tu, Brute?" Echoes "Eloi, Eloi Lama Sabachrthani?" *Psychiatry* 70 (1): 52-58.

[579] Deecke, *ibid.,* 406.

[580] Deecke, *ibid.,* 406.

[581] Deecke, *ibid.,* 411.

[582] Deecke, *ibid.,* 410.

[583] Jennifer, R. (2009). The Impact of Unconscious Communication. *Gallup Management Journal Online.* 1. Interview.

[584] Jennifer, *ibid.*

[585] Gore, A. (2014). *The Future.* New York: Random House: 40.

[586] Weber, J. M. (2007). Mirror neurons networks: Implications for modelling and consumer behaviour strategies. *Academy of Marketing Studies Journal.* 11, 2Z: 57-68.

[587] Gore, *ibid.,* 70.

[588] Brand, S. (1987). *The Media Lab: Inventing the Future at MIT*, New York: Viking Penguin: 202.

[589] Hendler, J., and Golbeck, J. (2008). Metcalfe's Law, Web 2.0, and the Semantic Web. *Web Semantics Science, Services and Agents on the World Wide Web*, 6 (1): 14-20.

[590] Halford, S., Pope, C., and Weal, M. (2012). Digital Futures? Sociological Challenges and Opportunities in the Emergent Semantic Web. *Sociology*, 47 (1): 174.

[591] Halford, Pope and Weal, *ibid.,* 179.

[592] Kristina, J. (2013). *Ten Digital Trends.* Retrieved from: http://trends.clickhere.com/the-evolving-social-media-landscape/.

[593] Chainese, A., Fasolino, A. R., Moscato, V., and Tramontana, P. (2011). A Novel Approach For Semantic Interoperability in the Web Based on the Semantic Triangle Communication Model. *International Journal of Software Engineering and Knowledge Engineering.* 21 (7): 1037.

[594] Chainese et al., *ibid.,* 1038.

[595] Chainse et al., *ibid,* 1038.

[596] Chainese et al., *ibid.,* 1040.

[597] Chainese et al., *ibid.,* 1045.

[598] Pollock, J. T. (2009). *Semantic Web For Dummies.* Hoboken: Wiley: 10.

[599] Deleuze, G and Guattri, F. (1987). *A Thousand Plateaus: Capitalism and Schizophrenia.* Translated by B. Massumi, Minneapolis: University of Minnesota Press: 26.

[600] Pollock, *ibid.,* 21.

[601] Geng, H., and Xu, S. (2011). Unconscious Self-processing: Subconscious, Unintentional or Subliminal? (Pp. 93-107). In Shihui, H and Poppel, E. *Cultural and Neural Frames of Cognition and Communication.* Dordrecht: Springer: 93.

[602] Choi, Y. S., Gray, H. M., and Ambady, N. (2004). The Glimpsed World: Unintended Communication and Unintended Perception. (Pp. 309-333). In Ran R. Hassin, *The New Unconscious*. New York: Oxford University Press: 309.

[603] Geng and Xu, *ibid.,* 95.

[604] Geng and Xu, *ibid.,* 95.

[605] Geng and Xu, *ibid.,*103.

[606] Geng and Xu, *ibid.,* 100.

[607] Chartrand, T. L., Maddux, W. W., and Lakin, J. L. (2004). Beyond the Perception-Behaviour Link: The Ubiquitous Utility and Motivational Moderators of Nonconsciousness Mimicry. (Pp. 353-354).In Ran R. Hassin, *The New Unconscious*. New York: Oxford University Press:

[608] Chartrand, Maddux and Lakin, *ibid.,* 345.

[609] Chartrand, Maddux and Lakin, *ibid.,* 357.

[610] Hatfield, E., and Cacioppo, J. T. (1994). *Emotional Contagion.* Cambridge: Cambridge University Press.

[611] Schachter, S. and Singer, J. (1962). Cognitive, Social, and Physiological Determinants of Emotional State. Psychological Review. 69: 379–399.

[612] Schachter and Singer, *ibid.*

[613] Schacter and Singer, *ibid.*

[614] Beattie, G., and Ellis, A. (2014). The Psychology of language and Communication. Hoboken, N. J. Taylor and Francis, 25.

[615]Sacks, O. (2012). *Hallucinations*. MacMillan, London: 60.

[616] Lera, G., Herrero, N., Gonzalez, J., Aguilar, E, Sanjuan, J., and Leal, C. (2011) Insight among Psychotic Patients With Auditory Hallucinations. *Journal of Clinical Psychology*. 67 (7): 705.

[617] Lera, et al., *ibid.,* 701.

[618] Sacks, *ibid.,* 67.

[619] Posey, T. B., and Losch, M. E. (1983). Auditory hallucinations of hearing voices in 375 normal subjects. *Imagination, Cognition, Personality*, 3: 99-113.

[620] Pierre, J. M. (2010). Hallucinations in Nonpsychotic Disorders: Toward a Differential Diagnosis of "Hearing Voices." *Harvard Review of Psychiatry*. 18:1. 23.

[621] Pierre, *ibid.,* 24.

[622] Pierre, *ibid.,* 27.

[623] Haddock, G., McCarron, J., Tarrier, N., and Faragher, E. B. (1999). Scales to measure dimensions of hallucinations and delusions: The Psychotic Symptom Rating Scales (PSYRATS). *Psychological Medicine,* 29: 879-889.

[624] Hunter, M. D. (2004). Locating Voices In Space: A Perceptual Model For Auditory Hallucinations? *Cognitive Neuropsychiatry*, 9 (1/2): 95-96.

[625] Singh, G., Sharan, P., and Kulhara, P. (2003). Phenomenology of hallucinations: A Factor analytic approach. *Psychiatry and Clinical Neuroscience,* 57: 333-336.

[626] Jaspers, K. (1997). *General Psychopathology*, translated by J. Hoenig, and M. W. Hamilton. Johns Hopkins University Press, Baltimore, MD.

[627] Langer, A. I., Cangas, A. J., Perez-Moreno, P. J., Carmona, J. A., and Gallego, J. (2010). Relation between Experiential Avoidance, Hallucination-Like Experiences and Clinical Symptoms, In A Non-Clinical Spanish Sample. *Journal of Cognitive and Behavioural Psychotherapies*. 10 (2): 162.

[628] Pixley, M. M. (2012). Some Difficulties with Psychodynamic Psychotherapy Attending to Benign Auditory and Visual Hallucinations. *Journal of Therapy Integration.* DOI: 10.1037/aoo29579: 3.

[629] Friston, K. J. (2005). Hallucinations and Perceptual Inference. *Behavioural Brain Science*, 28: 764-6.

[630] Lera et al., *ibid.* 1.

[631] Hugdail, K. (2009). 'Hearing voices': Auditory hallucinations as failure of top-down control of bottom-up perceptual processes. *Scandinavian Journal of Psychology*, 50, 542.

[632] Hunter, *ibid.,* 1010.

[633] Langer, et al., *ibid.*

[634] Hugdail, *ibid.,* 556.

[635] Pierre, *ibid.,* 29.

[636] Hugdail, *ibid.,* 556.

[637] Freud, S. (1914/1948). *Psychopathology of Everyday Life.* Introduction by A. A. Brill, Ernest Benn Limited, London: 192.

[638] Sacks, *ibid.,* 67.

[639] Pierre, *ibid.,* 24.

[640] Hugdail, *ibid.,* 553.

[641] Hugdail, *ibid.,* 553.

[642] Hugdail. *ibid.,* 553.

[643] Pierre, *ibid.,* 22.

[644] Barrett T.R. and Etheridge J.B. (1992) Verbal Hallucinations in Normals, 1; People who hear 'voices.' *Applied Cognitive Psychology* 6, 379–387.

[645] Pierre, *ibid.,* 25.

[646] Tien, A. Y. (1991). Distributions of hallucinations in the population. *Social Psychiatry and Psychiatric Epidemiology,* 26: 287-292.

[647] Pierre, *ibid.,* 22.

[648] Pierre, *ibid.,* 25.

[649] Pierre, *ibid.,* 25.

[650] Pierre, *ibid.,* 25-26.

[651] Pierre, *ibid.,* 26.

[652] Hill, K., Varses, F., Jackson, M., and Linden, D. E. J. (2012). The relationship between metacognitive beliefs, auditory hallucinations, and hallucination-related distress in clinical and non-clinical voice-hearers. *British Journal of Clinical Psychology*, 51: 442.

[653] Jaynes, J. (2000) [1976]. *The Origin of Consciousness in the Breakdown of the Bi-cameral Mind.* Houghton Mifflin.

[654] Pinker, *ibid.,* 77.

[655] Diederen, K. M. J., and Sommer, I. E. C. (2012). Eds. Auditory Verbal Hallucinations. *Hallucinations: Research and Practice.* Springer Science and Business Media, Dordrecht: 109.

[656] [657] Hugdail, *ibid.,* 553.

[657] Hugdail, *ibid.,* 554.

[658] Hugdail, *ibid.,* 554.

[659] Hugdail, *ibid.,* 554.

[660] Hugdail, *ibid.,* 554.

[661] Hunter, *ibid.,* 102.

[662] Hunter, *ibid.,* 102.

[663] Sugimori, E., Asai, T., and Tanno, Y. (2011). Sense of agency over thought: External misattribution of thought in a memory task and proneness to auditory hallucination. *Consciousness and Cognition.* 20: 689.

[664] Hunter, *ibid.,* 102.

[665] Sugimori, E., Asai, T., and Tanno, Y. (2011). Sense of agency over thought: External misattribution of thought in a memory task and proneness to auditory hallucination. *Consciousness and Cognition.* 20: 689.

[666] Sugimori, Asai, and Tanno, *ibid.,* 690.

[667] Hunter, *ibid.,* 97.

[668] Jaspers, *ibid.*
[669] Hunter, *ibid.,* 97.
[670] Hunter, *ibid.*
[671] Nazimek, J. M., Hunter, M. D., and Woodruff, P. W. R. (2012). Auditory hallucinations: Expectation-perception model. *Medical Hypotheses*, 78: 802.
[672] Nazimek, et al., *ibid.,* 801.
[673] Kompus, K., Westerhausen, R., and Hugdail, K. (2011). The 'paradoxical' engagement of the primary auditory cortex in patients with auditory verbal hallucinations: A meta-analysis of functional neuroimaging studies. *Neuropschologia.* 49: 3361.
[674] Pierre, *ibid.,* 28.
[675] Offen, L., Waller, G., and Thomas, G. (2003). Is reported childhood sexual abuse associated with the psychopathological characteristics of patients who experience auditory hallucinations? *Child Abuse*, 27: 919-27.
[676] Pierre, *ibid.,* 31.
[677] Freud, S. (1891/1953). *On Aphasia.* International Universities Press, London: 44.
[678] Langer, et al., *ibid.,* 162.
[679] Langer, et al., *ibid.,* 162.
[680] Pierre, *ibid.* 31.
[681] Lera et al., *ibid.* 9.
[682] Pierre, *ibid.,* 22.
[683] Pixley, *ibid.,* 4.
[684] Pierre, *ibid.,* 22.
[685] Barrowcliff, A. L. and Haddock, G. (2010). Factors affecting compliance and resistance to auditory command hallucinations: perceptions of a clinical population. *Journal of Mental Health.* 19 (6): 542-555.
[686] Barrowcliff, *ibid.,* 544.
[687] Pixley, *ibid.,* 4.
[688] Norris, F. H., Steven, S. P., Pfefferbaum, B., Wyche, K. E., and Pfefferbaum, R. L. (2008). Community Resilience as a Metaphor, Theory, Set of Capacities, and Strategy for Disaster Readiness. *American Journal of Community Psychology,* 41 (1): 127-150.
[689] Longstaff, P. H. and Yang, S. U. (2008). 'Communication Management and Trust: Their Role in Building Resilience to "Surprises" Such as Natural Disasters, Pandemic Flu, and Terrorism.' *Ecology and Society,*

13 (1). Retrieved from: http://www.ecologyandsociety.org/ vol13/iss1/art3/: 1.
[690] Norris, et al., *ibid.*
[691] Longstaff and Yang, *ibid.*
[692] Ross, H., and Carter, R. W. (2011). Natural disasters and community resilience. *Australasian Journal of Environmental Management*. 18: 1: 2.
[693] Kerr, P. (2013). Human Security in A. Collins. (Ed.). *Contemporary Security Studies*. Oxford: Oxford University Press.
[694] Peterson, C., and Park, N. (*2006*). A positive psychology perspective on post 9/11 security. *Basic and Applied Social Psychology*, 28, 357-361.
[695] Peterson and Park, *ibid.,* 357-361.
[696] Ein-Dor, T., Mikulciner, M., and Shaver, P R. (2011). Attachment insecurities and theprocessing of threat-related information: Studying the schemas involved in insecure people's coping strategies. *Journal of Personality and Social Psychology,101*(1), 78-93.
[697] Bowlby, J. (1988). *A Secure Base: Parent-Child Attachment and Healthy Human Development.* London: Routledge.
[698] Ein-Dor, et al., *ibid.,* 78-93.
[699] Bowlby, *ibid.*
[700] Ein-Dor et al., *ibid.*
[701] Ein-Dor et al., *ibid.*
[702] Mikulincer, M., Shaver, P.R. & Pereg, D. Motivation and Emotion (2003) 27: 77. doi:10.1023/A:1024515519160
[703] Ein-Dor et al., *ibid.*
[704] Ein-Dor et al., *ibid.*
[705] Ein-Dor et al., *ibid.* 79.
[706] Baldwin, M. W. (2007). Commentaries: on priming security and insecurity. *Psychological Inquiry*, *18*(3): 157-196.
[707] Baldwin, *ibid.,* 157-196.
[708] Baldwin, *ibid.,* 15.
[709] Baldwin, *ibid.,* 157-196.
[710] Harries, T. (2008). Feeling secure or being secure? Why it can seem better not to protectyourself against a natural hazard. *Health, Risk and Society, 10* (5): 479–490.
[711] Harries, *ibid.,* 479-490.
[712] Harare, *ibid.,* 9.
[713] Harare, *ibid.,* 11-12.
[714] Mossman, *ibid.,* 214.

[715] Nelson, B. (2012, May 2). *Projecting Human Evolution: 5 Traits We Might Possess in the Future. Earth Matters.* Retrieved from: http://www.mnn.com/earth-matters/wilderness-resources/stories/projecting-human-evolution-5-traits-we-might-possess-in.

[716] Fauen, E. (2012, November 26). *Top 10 Possible Next Steps in Human Evolution. ListVerse.* Retrieved from: http://listverse.com/2012/11/26/top-10-possible-next-steps-in-human-evolution/.

[717] Fauen, *ibid.*

[718] Nelson, *ibid.*

[719] Zolfagharifard, E. (2014, September 12). Are we evolving into a NEW TYPE of human? Different species will have evolved by 2050, scientist claims. *MailOnline.* Retrieved from:http://www.dailymail.co.uk/sciencetech/article-2752166/Are-evolving-NEW-type-human-Different-species-evolved-2050-scientist-says.html.

AUTHOR CONTACT INFORMATION

Dr. Luke Strongman
Social Sciences
Open Polytechnic of New Zealand
Email: Luke.Strongman@openpolytechnic.ac.nz

INDEX

M

N

P

R

S

T

U

Z